GENOGRAMS

The New Tool for Exploring the Personality, Career, and Love Patterns You Inherit

EMILY MARLIN

CB

CONTEMPORARY BOOKS

Library of Congress Cataloging-in-Publication Data

Marlin, Emily.
    Genograms : the new tool for exploring the personality,
career, and love patterns you inherit / Emily Martin.
        p.   cm.

    T.p. verso.
      Includes index.
      ISBN 0-8092-4494-2
      1. Personality.   2. Life style.   3. Family—Psychological
aspects.   4. Genetic psychology.   5. Genealogy.   I. Title.
BF698.M329   1989
155.9′24—dc20   89-9717
                                            CIP

In memory of Ivie Maud Miller

The genogram on page 2 was originally published in *Self* magazine.

Published by Contemporary Books
An imprint of NTC/Contemporary Publishing Company
4255 West Touhy Avenue, Lincolnwood (Chicago), Illinois 60646-1975 U.S.A.
Manufactured in the United States of America
International Standard Book Number: 0-8092-4494-2
20   19   18   17   16   15   14   13   12   11   10

CONTENTS

ACKNOWLEDGMENTS

Growing up in a small Pennsylvania town, I was lucky to have many relatives in my life. My father is one of eight children, and my mother is one of ten. Both families lived, and pretty much continue to live, within a few miles of each other. Family was, and still is, very important to me.

My grandparents were an enormous influence on me. I was always fascinated by the family stories they had to tell, and by their photographs that told stories about earlier times. One of my fondest memories of childhood is of the Miller reunion. Each summer there was a big family picnic in Chester County, where my paternal grandmother's family gathered. The highlight of the reunion was my grandmother's reading her "budget." These were humorous incidents about particular folks in the family, which she collected throughout the year. When my grandmother got up to read her budget, you could be sure that everyone listened. She was a great storyteller. She had a beautiful voice, a twinkle in her eye, and a marvelous wit.

There is no way I feel I can ever compare myself with my grandmother. But in completing this book, I'd like to think I am somehow following her tradition of collecting and telling family stories. But this time, I've done it from a family-therapy perspective in the form of genograms.

The founder of Bowen family systems therapy, Murray Bowen, M.D., pioneered the genogram concept and the systems theory which are the intellectual basis for this book. Equally influential, and the basis for my understanding and application of genogram construction, is Monica McGoldrick and Randy Gerson's definitive masterwork, *Genograms in Family Assessment* (New York: W. W. Norton, 1985). McGoldrick's dedication to the standardization of genogram format and her comprehensive explanation of the principles underlying its interpretation are unsurpassed. Her work has resulted in genogram consciousness-raising for thousands of therapists.

In the interest of simplicity, I have strayed somewhat from the standardization set forth by McGoldrick and Gerson. In no way do my variations on the theme imply theoretical departure; I merely intend to make genograms accessible to the lay beginner. I hope this book encourages readers to learn more about their families, and to seek out the works of Murray Bowen, Monica McGoldrick, Betty Carter, and other distinguished colleagues in the field of family therapy.

I also owe many thanks to the people who shared their family genograms with me and permitted me to report their stories and insights. To protect my sources, I have, of course, changed names, dates, and other identifying information and a few genograms are constructed from composites. I'm very grateful to my clients and friends for giving me their fascinating histories and to my kin for giving me mine.

INTRODUCTION

Have you ever said, "That's the story of my life!" referring to similar scenarios that seem to get played out in your life over and over again? Everyone loves a good story. Good stories have intrigue, colorful characters, and fascinating plots. Have you ever thought about your family history as a good story? It probably has all of the ingredients of an excellent book

The genogram, or family tree, is a useful diagram of your family history that can remind you of the many stories in your family. It can point to other scenarios that you may not have previously considered vital to your family history and your place in the world. The genogram can show you that the story (or stories) of your life may not be all that different from the stories of other people in your family. Strikingly similar repetitions in other people's lives may pop out at you as you draw a diagram of your family tree.

The process of drawing in the people, events, and relationships in a genogram is certain to reveal a variety of themes and patterns that have characterized your family over the years. It not only documents what events happened in your family, but gives you clues about why they happened and why they happened at particular points in the family history. These old family stories will be spelled out on the genogram in such a way that you will observe connections between what happened in your family and what is happening in your life right now. These messages make the genogram a family tree that talks.

Have you ever caught yourself saying things like, "my family always . . . ," "we [the family] do it this way . . . ," "our family never . . . ," "right or wrong, the family . . . "? Families, to a large extent, prescribe what we do, think, and say. We may like to think of ourselves as mavericks in our family stories, but most of the time we are simply being copycats. The genogram helps us see just how alike and different we really are from our families.

People often say things like, "I sound more like my father every day," "I thought it would never happen to me because of what I saw in my own family," "I hate it when I find myself sounding just like my grandmother: nag, nag, nag." Do any of these remarks sound vaguely familiar? It's not at all surprising that we become, in part, the parents we swore we'd never be.

Parents tell us to do as they say, but we do as they do. Our families are definitely one of the most important influences in our lives. From our parents and other family members, we learn expectations and patterns of behavior that shape us and all of our relationships. Following these patterns and living up to expectations helps establish who we are, what we believe, and how we behave at work, at home, and in our social circles. Despite all our attempts to flee (by rebellion or geographical changes), we tend to follow our families. Our destiny is often our history.

The fact that history repeats itself is not necessarily bad news. Those "old sweet songs" we know so well are sometimes sweet and sometimes bittersweet. For instance, you might feel proud of the strong traditions and moral fiber of your family history. You might feel good about upholding these standards and passing them on to your children. Your family stories might beef up your self-esteem and give you a sense of confidence and accomplishment.

However, on the bittersweet note, you may not be particularly happy with the way old family directions are not in sync with the family directions of the person you love. Divergent family rules and practices can create conflict between two people from different families. Doing the genogram may show you that conflicts about making and spending money, raising children, taking vacations, and so forth have to be negotiated in a present relationship because each person received very different directions from his or her respective family.

The genogram describes your family "inheritances," and some of your family gifts may not please you. You might even want to reject some of these inheritances and alter history's cycle. The genogram gives you the right information so you don't have to keep making wrong choices. It can help you make the right choices for yourself. It can be a wonderful facilitator for change. Seeing the influences can help you evaluate their effects and identify what you can do to make changes in your own life.

Thus, plotting a genogram links old patterns with what is going on in your life today. It is a fascinating way to unravel relationship patterns. It pinpoints unique and special family characteristics. It can chart your psychology as well as your history, feelings as well as facts.

This book explains the magic and mystery of family life—*your* family life. It can be a kind of trip down memory lane. It is the opportunity for you to draw a family tree that can come alive with an abundance of interesting facts and folklore. It will not only graphically document people, places, and events (like births, deaths, marriages, divorces, tragedies, and great love affairs), but it will show how your family lived and how your family lives in you.

Get ready for an exciting journey. This book will be your guide every step of the way. As you read about the genogram process and learn to draw your family tree, you will have a wonderful time. Enjoy digging into your family roots.

1
GENOGRAMS: FAMILY TREES THAT TALK

What Is a Genogram?

A genogram is a type of family tree, a diagram of a family over three generations. It follows the basic format shown in Diagram 1 to depict blood and marriage relationships from your grandparents' generation to your own. What takes this diagram beyond the standard family tree is that you enter information about ages, occupations, behavior patterns, and other observations that will help you understand your family better.

Genograms can be extended to include family stories or myths, whose themes may still be present in your generation. You may be surprised to learn how these stories have shaped your life. They play a big part in establishing your identity. Closer attention to the stories will suggest that they are more than anecdotal or amusing tales. They carry instructive messages that tell people in the family who they are and how they should behave. You are who you are partially because your family told you its expectations.

DIAGRAM 1 — ALONE AGAIN?

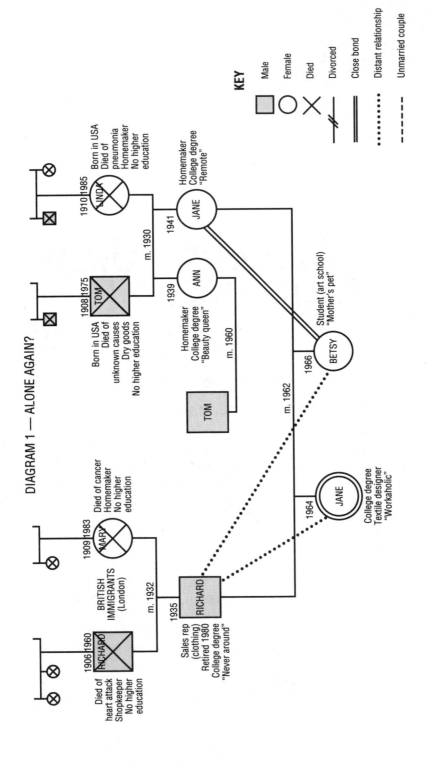

KEY

⬛	Male
○	Female
✕	Died
╫	Divorced
‖	Close bond
⋯	Distant relationship
╌	Unmarried couple

BRITISH IMMIGRANTS (London)

RICHARD 1906–1960
Died of heart attack
Shopkeeper
No higher education

MARY 1909–1983
Died of cancer
Homemaker
No higher education

m. 1932

TOM 1908–1975
Born in USA
Died of unknown causes
Dry goods
No higher education

LINDA 1910–1985
Born in USA
Died of pneumonia
Homemaker
No higher education

m. 1930

RICHARD
Sales rep (clothing)
Retired 1980
College degree
"Never around"

1935

ANN 1939
Homemaker
College degree
"Beauty queen"

JANE 1941
Homemaker
College degree
"Remote"

m. 1960

TOM

m. 1962

m. 1964

JANE
College degree
Textile designer
"Workaholic"

BETSY 1966
Student (art school)
"Mother's pet"

Your genogram puts *you* on your genogram in a very significant way. It graphically shows where you are in relation to others in the family. It clearly identifies where and how you are an extension of many old family roots. The genogram diagrams emotional links as well as genealogical lines. You may learn you are more connected to your family than you thought.

Has anyone in your family ever said things like, "You look just like your Aunt Bessie," or, "You're acting just like Grandpa Jones," or, "Uncle Bill loved fly fishing, too"? Family trees are visual diagrams of many of these kinds of likenesses and differences.This book shows you how to diagram these relationships, suggests some ways to interpret them, and tells you how you can benefit from the process.

How Did Genograms Develop?

People have been keeping track of their families for centuries. Even before written words were set down, family legend was passed down in family lore. Storytelling was a valuable means of passing along the family history. Once people learned to read and write, births, marriages, and deaths were frequently recorded in family Bibles. Today some people prepare complicated genealogical family trees. Most families have kept some record of family descent. Whether families tell stories or write their histories, family roots are "planted" somewhere. They survive because they've been carefully sowed and nurtured.

Genealogical societies and historical groups have long used family trees to account for family pedigrees. But it has only been in recent years that family physicians and family therapists have developed the genogram as a form for assessing family functioning. The historical family tree has become a more physiological and psychological family tree. Physicians and other clinicians have found ways to adapt the traditional family tree to chart a client's medical or psychological history.

Murray Bowen, M.D., is credited with developing the genogram in the late 1970s as a tool for analyzing family structures in his family systems theory. More recently, family therapists Betty Carter, Monica McGoldrick, and Randy Gerson have refined Bowen's model in their attempt to standardize the genogram to facilitate its use by many professionals who work with families. An increasing number of family therapists, physicians, nurses, and others are using the genogram form and symbols, along with the principles recommended by Bowen, Carter, McGoldrick, Gerson, and others, to broaden their perspectives on the individuals and families with whom they work.

Why Do a Genogram?

According to the American Genealogical Research Institute, "Genealogy has been likened to a gigantic jigsaw puzzle . . . a jigsaw puzzle in which you yourself are one of the pieces. It is that, certainly, but at the same time it is bigger than that. It is a puzzle with life breathed into it. Life and, yes, excitement. It is nothing less than a hunt with all the elements of a game. A hunt in which you must use your ingenuity and perseverance. A hunt which may ultimately lead you halfway around the world on paper or in actuality, and yet, almost paradoxically, back to yourself and a greater realization of your own place in the finely woven fabric of the history of mankind."*

You can benefit from the process of doing a genogram as well as from what it shows you. Depending on how you approach the activity, it can be a simple way to entertain yourself or a more complex untangling of family issues.

It's Interesting and Fun

The most basic reason for preparing a genogram is that it's interesting and fun to do. The genogram is a fascinating script, a drama of your family, that describes how you live your life. Your family tree is undoubtedly much more interesting and intriguing than most of the books you read or films you see. Truth *is* stranger than fiction, and your

---

*American Genealogical Research Institute Staff, *How to Trace Your Family Tree.* (New York: Dolphin Books, Doubleday & Co., Inc., 1974), p. 6.

family history is rich in tales of love and war, mystery and adventure.

What makes this drama so interesting is that we're players in it. Deeply ingrained in all of us are rules of human standards of social interaction, belief systems, and certain expectations. These have been passed down to us in "scripts" from our predecessors. Families have different scripts, and in any one family there may be several sets of directions. Usually these scripts strengthen family life. They cast our families in a very appealing light; they make them brave, passionate, and successful. But sometimes they portray a dark or sinister view of family life, full of anger, suspicions, or resentments.

Betsy, a thirty-five-year-old part-time student, uncovered the script for women in her family. She spends a great deal of her time managing her husband's life. Doing her family genogram made her realize how closely this behavior follows her family's script that says, "You have to take care of men; they cannot take care of themselves. The only person you can rely on is yourself."

"My dad had a nervous breakdown when I was in high school," says Betsy, "and my mother supported us after that. He never did go back to work and mostly hung around the town doing odd jobs for the neighbors. Even before that, though, my mother considered him a 'failure.' He had worked for my mother's father, a very successful self-made man, so I don't think we ever thought of my father as having a 'real' job. Actually, he is an extremely intelligent, sensitive man. But we all saw him as a person you couldn't depend on, a person who needed a caretaker or fixer.

"The other man in our family is my brother, the 'baby' of the family. And that word pretty well sums up how we see him—the helpless child. He is thirty-three, still lives at home with mother, and is trying to figure out what he wants to be when he grows up. Meanwhile, my two sisters have married men who are not nearly as successful in the business world as they are.

"I definitely think the women in the family got the message that you cannot rely on men for anything. That's my mother's script, and we learned it quite well."

If you enjoy mystery stories, you're sure to have fun with your genogram. The process of drawing your genogram may give you some clues about family secrets, describe the black sheep, and unlock the skeletons in the closet. You can get more clues by conducting personal interviews. Investigating your family in a nonthreatening conversation (not an interrogation) can lead to the sharing of secrets. You can say you're collecting the facts for your family tree. The people you interview for family information may clear up some old family myths or mysteries. They might tell you of plots you never knew about or scenes you never witnessed.

Frannie, a twenty-seven-year-old film editor, had such an experience. On a recent trip back home, 2,000 miles from where she now lives, she had a long conversation with her grandmother and unraveled some old family mysteries.

"I often wondered why my family never spoke about religion or attended church," she says. "When my grandfather died six years ago, no one seemed to know exactly what to do to make any kind of funeral service. My mother didn't want any clergy in attendance, and it was a very strange and uncomfortable scene. No music, no prayers, just a couple of members of the family saying their good-byes out loud.

"My grandmother told me about her life, which no one had ever spoken about before. Her mother had been a Roman Catholic nun and had left the convent in her twenties. She married my great-grandfather and went on to have thirteen children; my grandmother was her youngest. My great-grandmother died in childbirth when my grandmother was two years old. Later, Grandmother was adopted by a couple who never told her of the adoption. Somehow, though, my grandmother knew but never let her adopted parents discover she knew. She loved her adopted parents very much and really wanted to protect them.

"Her adopted parents were very religious Catholics, and they disowned my grandmother when she married my grandfather, a Protestant. They refused to have anything to do with her after that. My grandmother was very hurt, and once again she lost her parents. She never again went to the Catholic church or any church.

"My mother married my father, who was an atheist, so we didn't go to church either. We grew up without any kind of religious indoctrination. Sometimes I think anything would have been better than nothing. Religion continues to be a very mysterious subject in our family. My parents felt their children should choose their own religion. But that was a little hard to do, since we didn't have any frame of reference at all. I often wondered why religion was a taboo subject. At least, now the mystery is cleared up about why no one in the family talked about religion or went to church. It was a very painful subject in the family."

The genogram chronicles the fact that your family story has more intriguing plots than your favorite soap opera, more tragedy than historical novels, and more surprises, romance, and comedy than the theater, opera, or circus. Those old stories in your family may be big news to you. The genogram helps you to "read all about it."

Carole, a thirty-five-year-old property manager in business with her husband, learned of a terrible tragedy in her husband's family. A cousin was instantly killed in an automobile accident where the twelve-year-old daughter, Judy, was a passenger in the back seat. The child was physically unharmed, but she witnessed her mother's death. Understandably, Carole's heart went out to the child and she asked, "How does a child ever recover from the death of a parent?"

In studying her genogram, Carole saw that this terrible news was very closely related to old news about her family. Her mother had also lost her father when she was twelve. Carole began to understand that perhaps her mother had never fully recovered from that profound loss.

"Looking at the genogram," says Carole, "I see that not only was my mother the same age as Judy, but my mother was left in a very unprotected position in her family. Her mother had to care for three children, a mother-in-law, and another elderly woman who lived with them. I don't think my mother's mother had time to give her youngest child much attention at that time. And family stories tell about what 'hell-raisers' my mother's twin older brothers (by five years) were. She bore the brunt of many of their pranks, which probably were closer to neglect or abuse than pranks. My mother never liked her brothers but didn't explain why. She was her father's favorite, and I guess after he died, she didn't really have anyone who could protect her from the 'hateful' twins."

It Will Help You Understand Your Family

As Carole found, genograms are not only fun, but they can serve as your introduction to the history of your family, which you can, if you like, pass on to your children and future generations. Doing the genogram gives you the opportunity to acknowledge the many strengths and important truths of your family. You will get new perspectives on things you took for granted, discounted, or never saw. As you get closer to your family roots, you may find they are much more nurturing and supportive than you ever imagined. The nourishment of understanding and empathy will help your family tree grow and flourish.

Doing the genogram is a rare opportunity to view your family in a new light. Your revelations will probably include innumerable strengths, courageous acts, and long traditions of loyalty and love. You will also find secrets, problems, and pain. Family weaknesses are not flaws; they make you and your family vulnerable and real.

As you discover your family's trials and tribulations, be prepared for some negative as well as positive emotional experiences. Doing a genogram often breaks through denial and points to family "troubles." As you start to identify with the more painful experiences of your family, you will feel the pain. But your identifications with family struggles and sorrows will make you feel caring and sympathetic toward yourself and your relatives. The point of doing your genogram is to affirm yourself within the context of your family history. The process starts with family examination and ends with self-discovery and self-validation.

Your genograms will give you insight into how normal or dysfunctional your family is. Most families have their share of problems (conditions or circumstances like alcohol abuse, physical illness, violence) that makes them behave in a certain predictable fashion. Everyone in the family focuses on the problem and tries to keep the family intact in spite of a major disruptive factor that causes everyone emotional and/or physical pain. The family organizes itself around the dysfunction to keep the family going. The genogram can weigh the issue and show how people react to particular family problems.

Often when there is a problem, people either run from it or get entwined in trying to solve it. Family therapists call this being cut off or enmeshed. Maybe you are too close to (enmeshed in) or too distant (cut off) from your family. There are ways to find that comfortable balance so that you can stay connected without fear of being either too close or too far.

Kate, a forty-one-year-old restaurant manager, found that drawing her genogram gave her some perspective on the scale of how close or distant the members of her family were. She was trying to balance her life so that she wasn't overinvolved with her mother's need for protection from her father's violent temper or underinvolved with her own life.

"My father's drinking problem split the family in two," she explains. "My two oldest brothers took off and moved hundreds of miles away, and my sister and I stayed close by and tried to save them, especially my mother, who we felt we had to protect.

"Jan, my sister, is forty, has never married, and still lives at home. I moved out when I was twenty-three but stayed in the general area. I still call home every day and go there almost every week.

"In our family it was OK for the men to leave, but not the women. I guess men were expected to make their way in the world. They were the deserters and the women were rescuers. My sister and I still feel super-responsible and try to see that things at home don't get too out of hand. We, like most women on our genogram, are definitely rescuers. My mother's mother lived with us until she died at seventy-six, and my father's sister took care of her widowed father for fifteen years.

"This summer my sister and I plan to go to Europe for three weeks. This is the first time neither of us will be around. We've told our parents and brothers our plans, and we are just leaving it up to them to make their own arrangements. We might feel guilty, but we have to do it ourselves. I'm looking to find that balance where I feel I'm a dutiful daughter and still looking out for myself. I don't want to be a deserter or rescuer. Our parents haven't killed each other in forty-five years, and they probably won't when we're on the trip."

As Kate found, the genogram looks at you and reflects back to your family. It shows similarities and differences that you may have not seen before. Don't be surprised if the likenesses startle you; we tend to look or act very much like certain people in our families. Being named after a relative, for instance, almost guarantees certain mirror images (physically or behaviorally).

For example, Dee, a twenty-nine-year-old social worker, was named after an aunt with whom she strongly identifies. She explains how she is the mirror image of her namesake. "In doing my genogram I remembered how much I identify with my Aunt Dee, whom I'm named after. She was my mother's youngest sister, and my mother and her other sister really tormented Aunt Dee as a child. I recall stories about how the two older girls would run ahead of her, leaving her in the snow on the way home from school. They thought it great fun to 'lose' or 'forget' her. I got the impression from my mother that Aunt Dee was a 'royal pain in the neck' and 'a spoiled brat.'

"Of course, my mother always said I was 'just like' my aunt, and I believed it. I felt I was selfish and unwanted. I was a pain in the neck too. I don't remember my mother ever mentioning my aunt's name without rolling her eyes. I don't remember her rolling her eyes at me, but I certainly felt her disapproval all my life.

"Looking at my aunt's history on the genogram, I see she was a very sympathetic woman. Everyone came to her with their troubles. Maybe that explains why I was drawn to my profession as a social worker. I admired her and wanted to be like her. As a result, people confide in me, too.

"Even though my aunt was the target of ridicule in the family, she was an interesting and successful woman. She didn't get as involved in the family craziness as other people. She traveled all over the world and was a very together woman. I feel very close to her and am glad I was named after my aunt in spite of all the flak I got for being 'like' her."

As you gather information from relatives, your conversations can open up whole new areas of communication. The more questions you ask and the more answers you find, the less likely you are to harbor old fears, fantasies, and resentments. Talking, being open and honest, and showing interest and respect are good practices that can enhance all your relationships.

Throughout this book we will use the tree as metaphor. Your family tree has roots (solid or shallow), branches (strong or weak), and leaves (brilliant or dull). Your family has all of these characteristics and more. As there are many species of trees, there are many species of families. You will comb the forest to find out how your family tree is like other trees and how it is unique.

Drawing your family tree makes you realize how powerful your own family is. You discover what an impact these strong and sturdy roots have on you. For instance, you may see signs of extraordinary valor, noble deeds, high principles, and unusual resilience. The genogram acknowledges these and other worthy accomplishments and strong traditions. Seeing your family in a worthwhile perspective makes you realize what a strong family tree you have. You may better appreciate family strengths and increase your personal self-esteem as you admire and respect your heritage. Empowering your family means you can empower yourself.

While some roots in the family system are indeed powerful, others may show signs of weakness. It's important to acknowledge less-powerful roots also. Some family roots are blemished by terrible secrets, inglorious deeds, and serious problems. Blemished, however, is not blighted. Your family, like all families, is not perfect. All family trees have their imperfections: negative characteristics, flaws, and foibles. You can appreciate these vulnerable branches on your family tree when you recognize they are mixed in with all the solid ones.

It Will Help You Direct Your Life

Your genogram may uncover what appear to be coincidences. But these patterns of relationships and behavior are not coincidental; they are the result of learning family scripts. Watching the behavior of family members is the most fundamental way we learn how to behave ourselves. As we identify our family scripts (on the genogram), we can, if we wish, edit them. History doesn't have to be a repeat performance. We don't have to follow the same old script simply because it's the family script.

If, for instance, you and your partner have a conflict about when and how to spend money, you might look at your respective genograms to try to determine the value of money in your families. You can find the answer that works best in your relationship. It might be a compromise or combination of the two money styles you adopted from your families.

Some old family scripts closely resemble current scenarios. Displays of affection in one family may be met with disapproval, resistance, or anger by another family one enters by marriage. The second family may have a much cooler emotional tone and its members may feel quite uncomfortable with displays of affection.

Lillian, for example, wants to find family patterns that are causing problems in her marriage. Although she has been married only six months, she realizes that she has to practice changing behavior that she learned in her family.

"I am having a difficult time in my new marriage," she explains. "My husband, Bill, is a very warm, affectionate, and demonstrative person. Obviously, that's one of the things that attracted me to him. But sometimes I really panic when he grabs me and catches me off guard.

"Last night we were fooling around, and he pulled me onto his lap and wouldn't let me up. He was being playful, but I acted like a caged animal. I was in an absolute panic. It was as if I was not going to be let loose, and I had to break away.

"These frightened feelings remind me of what it was like as a kid when my mother would be drunk and very smothering. She would do things like that, hold me really really tight. I couldn't breathe, I couldn't release myself, and I was frightened and repulsed. It was just awful.

"I got really annoyed with Bill, but I haven't explained yet why I had such a horrible reaction to him. I overreacted, certainly, and I will try to explain why. At the time, I was just too upset.

"But more than that, I have to try to break the habit of reacting so strongly to his loving attention. He isn't my mother, and he isn't a slobbering drunk. I hope I can learn not to go into automatic panic when he's just being playful. If I could relax instead of violently reacting at those times, I might be able to enjoy his affection. But I think it's going to take more practice."

Lillian's reaction was understandable in light of her history. In this case, her husband's affection, which was playful in his family context, was painful to Lillian. It reminded her of her mother's exaggerated intimidating affection.

The genogram is a road map of your family. It will show you where you came from and where you are going. If you study the road map, you'll get to know the "territory"—the people, places, and events that make up your history. You might want to change your direction at some point and get on roads that are going to take you where *you* want to go and not to the same old places. As certain patterns of obstacles emerge on the road map, you may want to resurvey the landscape of your family background and choose a different route.

If people in your family were on the road to unhappiness or self-destruction (through, say, substance abuse or victimization), you probably want to get off that road if you are on it. If you have not yet come to a dangerous route but the going seems a bit rocky, you may want to chart a safer and better road.

Florence is one person who did this. She loves her new job in chiropractic medicine, but she feels somewhat guilty because she has stopped following the family career path (hated jobs) that leads to nowhere (except misery).

Florence's genogram showed that most of the "work traditions" on her family tree resulted in victimization. She no longer wanted to be a victim. Some of her guilt was from abandoning the family route; she didn't want to continue in that direction of being overworked, underpaid, and unhappy.

"It's almost a crime in our family to do something you like," she explains. "It's an even bigger crime to get well paid for it. My parents hated their jobs; they were both schoolteachers. Of course, they never considered making any changes. They constantly complained about how hard they worked, how unappreciated they were by the students and administration. And they never, obviously, got paid enough. My father taught summer school every single year to 'make ends meet.' He never took a vacation himself. The theme was, 'The road to success is closed.'

"My grandparents on my father's side had a mom-and-pop candy store and literally worked themselves to death. My grandfather died the week after he retired. My mother's parents were first-generation Americans; they never learned the English language. They were always disadvantaged, poor, and isolated. Their children supported them most of their lives, and they thought of themselves as failures.

"Much to my family's chagrin, I left my job as a nurse three years ago. I finally have a profession I love, and I expect to make a very good living as a chiropractor. It certainly goes against the family grain, but it sure feels good to me. I'm on the road to success, and I plan to stay on it."

Constructing your family tree will encourage you to think about your family life in new ways. You will uncover "inherited" talents, values, and ways of coping that can be but don't have to be irrevocable legacies. Chart your past with a view toward the future. The past does not *have* to be tomorrow's prologue. Today you can reject inheritances and break traditions. The genogram can be an instrument of change. Use it to make the changes you want to make.

How to Do a Genogram

While family therapists advocate the use of the genogram in professional family assessment, little has been written about how individuals can apply genograms to their own family explorations. There is no reason why you cannot incorporate the teachings of family therapists, historians, and genealogists to draw your own family tree. You can use the genogram format shown at the beginning of this chapter to chart your family history and the influences of your heredity.

The next two chapters will show you the procedure for adding information to the basic format. Along with the format and key of symbols, the book provides examples of family trees that have been drawn by other people, which you can use as a frame of reference. Later chapters then provide examples of some kinds of patterns you may want to look for.

Your genogram can be fairly simple or extremely complex, depending on how much information you decide to gather. But since there are so many details you could include, you should have some idea of why you want to do a genogram. This will help you focus on those details that are currently the most interesting or important to you. In the construction of your family tree, you have the opportunity to set your own agenda regarding what you want to learn from this experience. Each person has his or her own concerns and questions about the molding of the family and how it shapes the individual. You may be looking for information in father-daughter relationships in your family, or searching for an explanation for sadness, anger, or loneliness. You can program your genogram for that specific information.

You can track many things on your own genogram: feelings, attitudes, belief systems, social patterns, values, and behavior patterns, along with the more obvious names, dates, and events. This book provides ideas for selecting the material you want to investigate on your genogram.

You can make your family tree (genogram) a talking tree. The talking tree will not cure you of any big problems in your life, but it can start you thinking about the great influence your family has on you. It can pinpoint history and predict destiny. The genogram promotes both family reflection and self-discovery.

The exercise of constructing your genogram should be an emotional experience, not a mechanical drawing. It should be challenging, intriguing, and fun. No one is going to look over your shoulder or grade you on performance. It's strictly your exercise. Doing the genogram should be like a pleasant walk in the woods to find your family tree. This book will give you some suggestions about how you might go, but the walk is basically of your choosing.

2
GETTING STARTED

When to Do a Genogram

There is probably no perfect time to draw a genogram. There is just the time that seems best for you. It might be when your curiosity is piqued or when a personal problem is most painful. You may do the genogram exercise in a period of calm or a period of crisis.

During a Calm Period

Suppose that you are feeling fairly cool, calm, and collected about yourself and your family. You have a genuine healthy curiosity about your family and decide you simply want to know more about your beginnings. Your relaxed attitude will, undoubtedly, help you to draw a fairly objective family tree. Most likely you won't be charging your genogram with unresolved emotional issues.

Betty, a thirty-seven-year-old accountant in a brokerage house, did her genogram when she was feeling very good about herself. She had just received a promotion on her job and was contemplating the birth of her second child with much joy.

Betty's curiosity about her family was aroused after she received a long letter from her sister Suzanne (the family "historian") asking for some particular old family photographs. In looking through these pictures, Betty realized there was a great deal she did not know about certain family members. She decided to do a genogram so that she could have a record of her family history. She wanted to take advantage of the fact that she could ask Suzanne about the unknowns. Some day her own children might ask her questions about her family, and she wanted to have the answers.

But don't be surprised or alarmed if, once into the exercise, you begin to have emotional reactions. Repressed feelings of disappointment about old events or angry feelings about people you may not have thought about for some time may surface. Remember, they are feelings and not facts. And you can talk about these feelings with someone close to you.

During a Crisis

A crisis is seldom the best time to delve into matters that might make your crisis even bigger. However, plotting your genogram often relieves anxiety and offers information that can help you resolve conflict. It could give you new perspectives and offer possibilities for a positive change. A crisis might not be the worst time to draw your genogram.

Joe and Terri, a married couple in their forties, were having a big fight over a "lost" gold bracelet that was Joe's first anniversary gift, fifteen years ago, to Terri. Terri was angry at herself for not remembering where she had put the bracelet; Joe was angry and hurt that she could be so "careless" about something so significant. In addition, Terri was surprised at her own attitude of indifference. She loved Joe and knew it mattered a great deal to him, but somehow it didn't matter much to her.

Terri tried to understand why she was not more upset. She mapped out her genogram to learn whether her nonreaction was typical of how she had responded to previous losses in her life.

Terri's genogram showed that she had a pattern of burying painful feelings of loss and acting as if she were not devastated by losing something or someone very important to her. The first big loss that showed up on her genogram was the death of her first husband two and a half years after their marriage. Terri was twenty-four when Norman died of a cancerous brain tumor after a very brief illness. At that time, Terri threw herself into her college studies and became an A student. She buried her grief and never mourned Norman's death.

It was only when she drew her genogram that Terri realized it was exactly twenty-two years ago, almost to the day, that her husband had died. Since she had never allowed herself to grieve this profound loss, Terri stopped blaming herself for her nonreaction to the loss of the jewelry. She realized that by comparison, the material loss was much less significant than her prior loss.

Right Now

Following are some hypothetical situations that might prompt you to draw your family tree at this time. You may identify with some of these situations, or you may have your own personal agenda. You don't need a serious reason to draw your genogram. Personal curiosity is quite enough. It is helpful, however, to have some idea of areas and subjects that interest you. Some of the following cases may trigger your own reasons.

• Perhaps you recently read a book about a famous person's history over several generations. You loved the book and were fascinated by the complex legacies in one person's life. It made you think about how little you know about your own roots. You begin to wonder if your own family was a store of great adventures. You are curious and want to find out about your ancestors: where they were born, raised, and lived; what they did in their life's work; and how they fared in the world.

• Maybe you are having problems with your partner. A major complaint seems to be that you are "distant." You have heard this complaint before from other people. You decide to take it seriously and do some self-investigation. Since you cannot easily share your feelings, you decide to look at other people in your family to see if this is a family trait.

• Your daughter, the youngest of three, has just headed off for college, and you are depressed. You have read about the empty-nest syndrome but didn't think it could happen to you. You are flooded with painful feelings of loss and abandonment. You decide to draw a family tree to find out how family members dealt with their empty nests and other stress periods in the normal life cycle.

• Lately you are irritated with your friends who call you, asking for your attention and assistance. You feel overwhelmed and angry. Why, you ask yourself, are you once again in the caretaking role? You don't feel like the pillar of strength other people seem to think you are. A look at your family tree will show how you acquired this image.

Perhaps you will find out that you come from a family of fixer-uppers. Family tradition stipulates that you never say no to someone who needs help. You "should" be grateful that you are more fortunate so that you can help the less fortunate. This could be your family's message.

But isn't there such a thing as carrying a good thing too far? Do you always have to answer to everyone's beck and call? Perhaps your tree will tell you that you have a right to be needy, too. Maybe you will start to have more of a balance in your life, allowing yourself to get help as well as give it.

• Your best friend has been diagnosed with a fatal disease and has a month to six months to live. You love your friend and want to be supportive. But why then, you ask, are you enraged and frightened? You cannot believe that you are seriously thinking about taking a vacation to Europe or moving across the country.

Your genogram may reveal a pattern of tragedy and escape. In charting your genogram, you recall the story you heard for the first time when you were in college about your great-grandfather who went out for his nightly constitutional and never came home. He left behind a wife and ten children. Whenever you were sick as a kid, your grandmother was called in to take care of you because your mother "couldn't handle illness."

Perhaps you will be less critical of yourself when you

understand how difficult it has been for your family to deal with troubles and sickness. You have some choices, though. You don't have to follow this avoidance strategy; you can confront the normal feelings of anger and resentment that go along with sickness, death, and dying, perhaps finding a group for people in the same situation. You may find there is safety and comfort in numbers, especially a small number of people in a similar state. Simply being aware of your family's tendencies will help you to consciously seek the support you need.

• Once again you find yourself identifying with the women who "love too much." Your latest love just walked out on you, and even though he was not very nice, you want him back. Over and over again, you get dumped on and then dumped. Why, you wonder, can't you find a nice guy for once?

Your family tree may have something to say about abuse. Didn't your older brother pick on you unmercifully? He was the "little prince" and could do no wrong. You always got the blame for provoking him. No one seemed to notice what a bully he was, but you ended up feeling like the victim.

As you look closer at your family tree, you see there were other people who lived their lives in victim positions. Wasn't your Uncle Benny "poor Uncle Benny," the poor soul who was always taken advantage of by some cold-hearted woman? Your father frequently tells the tale of his "strict father" who sent him to his room without supper for the slightest disobedience of dining room manners. And then there was your alcoholic grandfather on your mother's side who tyrannized the whole family when he went on his weekly binges.

You certainly don't want to get caught in this abusive cycle. You deserve and need better treatment. You may need some help convincing yourself that you don't need to take it anymore, but by recognizing your family patterns, you have taken the first step.

These are a few hypothetical cases that may inspire you to chart your family history on the genogram form. Simple curiosity is also a powerful reason to do a genogram; you are sure to discover traits, tendencies, and patterns you never noticed before.

Where to Do a Genogram

Whether you do a very simple historical genogram or a more detailed tree noting personal characteristics and problems, you will want a place where you can concentrate without interruption. The genogram exercise requires that you systematically document the chronology of your family over three generations. It's best, obviously, not to pick a pressured time or place when you do this exercise. You can get the best results under the best conditions. An environment of peace and quiet is most conducive to making it an entertaining experience rather than a burden. You are creating a very special family tree, so try to choose a very special place to work.

What You Need

There are no wrong genograms or wrong ways to do genograms. You are a free agent in this exploration. To begin the procedure, you only need:

• Ordinary writing supplies
• The basic genogram form
• A key to symbols
• Instructions for each type of genogram

A graphic representation requires a graph, but your genogram doesn't have to be on graph paper. You can draw it on any large piece of paper. You will want plenty of space and an easy-to-read background. Black or dark blue illustrations on white paper show off words, symbols, and lines very well.

You might want to use a blackboard for your first draft so you can easily erase lines or change the dimensions of your genogram as needed. Then you can copy this draft onto a more permanent record.

Some people use big sheets of butcher paper and crayons. Others do a first try in pencil, using pen for the final analysis. They get down on the floor, kindergarten style, to draw their genograms. This creates a playful atmosphere.

You'll need writing material and writing implements. Pencils, pens, correction fluid or tape, eraser, extra paper, a ruler or straightedge, Scotch tape, and scissors, are tools of the trade in genogram construction. You can use colored pens or pencils to chart other characteristics beyond the basic names and birth, death, and marriage dates. You might, for instance, want to record nicknames in red and occupations in green. Think about it and decide what works best for you.

Putting generations on separate sheets of paper and then taping these sheets together is often less cumbersome than using one giant sheet of paper.

The next chapter will provide you with the basic format, keys to symbols, and instructions for the genogram types.

Once you are ready to begin, you need information about your family. There is probably a great deal you already know. For the rest, it's up to you to set your priorities, develop strategies, and gather information.

Information Gathering

The Primary Source: You

You are your primary source of information. Start out with what you know about your family. Before you start to draw your genogram, write down what you already know about your parents, grandparents, and siblings. Keep a date book of births, deaths, marriages, divorces, and other important family events. This log will be very helpful when you draw your family tree and will make an informative companion piece to the genogram. It will also point to information you do not have and will want to document later. Star or separately list any missing links.

For yourself and each of your parents, grandparents, and siblings, document the following information:

• Name

- Age
- Date of birth
- Place of birth
- If deceased, date and cause of death
- Occupation
- Education
- Religion
- If foreign born, date of arrival in this country
- Dates of marriage, separation, and divorce
- Names of children, sex, and birth dates in chronological order

This information gives you the basic demographics of your family. You want to know about all births, as well as miscarriages, stillbirths, and abortions. If people were adopted or are foster children, you will want to indicate this as well.

Documents

Once you have completed as much information as you know already, figure out how to go about obtaining the information you still want. These family documents might help you fill in some of the gaps:

- Family Bible
- Photographs
- Diaries, journals, and scrapbooks

For purposes of this genogram study, you probably won't want to do more than a first- or secondhand investigation. But these public documents, as well as advice from the American Genealogical Society, can be your guides for a more comprehensive research of your family history:

- Census data
- Property records

- Local court records
- Newspaper files
- Local libraries

Your Relatives

Your relatives are an excellent, if not 100 percent accurate, source of family information. Don't be too worried about vital statistics; you want to get ideas and impressions about your family, as well as cold facts. Even myths give a picture of family life. It will be interesting to compare descriptions or perceptions of different relatives about the same "facts," incidents, or situations.

There is probably one relative who knows the most about your family background. Most families have an unofficial and unappointed family archivist in every generation. He or she typically has a keen interest, a good memory, and an impressive collection of oral and written material about the origins of the family. Usually this person loves to impress other family members with his or her wealth of family history. The family archivist in your generation may be your first resource; maybe it's you. Ask your parents who in their generation knows the most about the family.

Don't rely exclusively on family archivists. Often the people you think will have the least to say have the most to say. That old saw, "You'll never know unless you ask," is very applicable to a family tree search. If you want to find out about the whole family, you have to ask the whole family. Different people will have different perspectives on the same events and dates. Emotionally significant events are most likely to account for discrepancies. Keep in mind, too, that memories are fallible. You are looking for the "spirit" of the family, so even inaccuracies and myths will tell you about your family's life.

Families are great networks, and once you announce your intention, you may start getting feedback from all kinds of people who want to tell you what they know. You might be presented with some golden opportunities to increase your knowledge of family and family process. A surprise phone call or visit from a distant relative might be just the time to get answers to those questions you have in your chronology notes or on your mind.

The Interviewing Process

Information gathering is usually easier than we imagine. Often we expect resistance that isn't even there. Generally speaking, people love to talk about themselves, and this includes experiences in their families.

How you approach your interviewees, however, is bound to have a great deal to do with how much information you elicit. You will improve your chances for success by being sensitive to the person or persons you are interviewing and by explaining that you are interested in finding out more about your family history. Explain what a genogram is and why you are drawing one. As you construct your family tree, show your interviewee what you are doing. That person may even want to do his or her own genogram later.

Naturally, you will want to choose a time and place that is as convenient and comfortable for your source as it is for yourself. If you start out by being courteous, genuinely interested, and open-minded about responses, you will probably get a wealth of information. Also, you'll get a very good sense of how your relatives fit into the family system.

You must anticipate and accept the fact that not every family member may be a willing subject. An individual may be too defensive or just plain unwilling to talk. Some people find it too dull or burdensome to talk about family history. They prefer to forget the past and live solely in the present. Respect the wishes of these people and either try later or go on to someone else who may be more receptive.

Questions you can ask relatives are the same questions you can ask yourself when you gather information for your genogram:

- Can you tell me where and when you were born?
- How old were your parents?
- When did they marry, or how old were they when

they married? How old were they when you were born?

- Do you have sisters and brothers?
- What is your birth order in the family?
- What are your siblings' names, and when were they born?
- What kind of work did your father do? Your mother?
- Where did you grow up?
- How far did you go in school?
- What kind of work do you do or have you done?
- What schooling did your siblings have?
- What are the jobs or careers of your siblings?
- When and whom did you and your siblings marry?
- Can you tell me the names and birth order of your children? What about those of your siblings?
- Were there any miscarriages, stillbirths, adoptions, or foster children in these families?
- Were there any separations or divorces?
- Can you tell me which of your relatives are deceased? Do you know when they died and the causes of death?

What do you know about your parents?

- Where were they born?
- Were their parents foreign-born? If so, where?
- When did they immigrate to this country? How old were they?
- What kind of work did they do?
- How far did they go in school?
- What was their religion?
- How did they meet?
- When did they marry?
- What was their birth order in their families?
- Can you give the names, dates of birth, occupations,

and educations of their siblings?

- If any of your parents or their siblings are deceased, can you tell me when they died and what their causes of death were?
- What do you know about the marriages and children of your parents and your parents' siblings?
- Their occupations and education?
- Were there any miscarriages, stillbirths, adoptions, or foster children in the families of your parents and their siblings?
- Were there any separations or divorces in those families?
- When, where, and under what circumstances did your parents' siblings die? And your parents?

The Importance of Family Research

Even if you think you know everything about your family, chances are you don't know it all. Most families have "secrets," subjects that are not talked about. There is usually an unspoken rule not to talk about certain family facts or problems. Physical and mental health problems are just two areas that are typically off-limits. Whether these illnesses are caused by genetic, environmental, or unknown causes, it would make good preventative sense for families to share these secrets.

At some point, research almost always involves breaking the "don't talk" rule. This rule isn't hard to break if you do it in the vein of drawing *your* family tree to learn about *yourself.* "Do you remember" is an easy way to open a conversation. If you take the initiative and relate a family incident, the family member to whom you are speaking will often counter with another recollection. You'll be amazed at how much you can find out once you have their interest. You may be surprised at all the opportunities for exploration that present themselves once you are on the lookout for family truths.

Often the family secrets come out in discussions like this: Jill and Jon, a brother and sister, are talking on the phone. Jill says, "Jon, do you remember how Aunt Joan always fell asleep at every festive family dinner?" Jon replies, "I remember that every Thanksgiving and Christmas, she would be out cold by the time dessert was served. Everyone totally ignored the fact that her head was in her mashed potatoes. We kids used to snicker with each other, but the adults just shushed us up and didn't say a word."

After this reply of Jon's, Jill wonders if it was simply that Aunt Joan had had too much egg nog on the holiday. She remembers that Aunt Joan left home in her twenties, and no one ever mentioned her name again. No one knew her whereabouts until years later when the family was notified of her death in a hotel fire. She had been living alone in a dingy hotel and apparently had fallen asleep with a lighted cigarette. It doesn't take much more research to hypothesize that Aunt Joan had a serious drinking problem.

Jon and Jill's conversation prompts them to look at the other family members in terms of alcohol abuse. This is where the genogram becomes important, for it gives meaning to the bare facts Jon and Jill turn up through research. The genogram (Diagram 2) alerts them to a pattern of abuse in every generation. They decide that some basic education about family alcoholism might be wise. Since they both have teenagers, they are particularly concerned about breaking the cycle of addiction in their family.

Some people love to do family research, and others have a more difficult time digging into their family roots. If you find your research tedious, difficult, or disturbing at some point, remember that the genogram will help give meaning to family facts and stories. What may at first seem to be a jumble of information and anecdotes can emerge into clear and meaningful patterns when you draw your genogram.

Bringing Family Trees to Life

Doing a genogram and gathering the research often resurrects fascinating family stories and themes. Here is

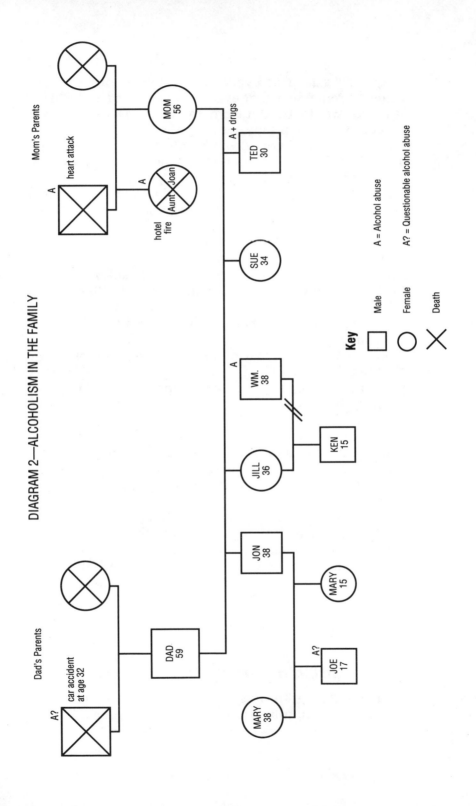

DIAGRAM 2—ALCOHOLISM IN THE FAMILY

Mom's Parents

heart attack

A

MOM
56

A + drugs

TED
30

hotel
fire

Aunt Joan

A

SUE
34

A

WM.
38

JILL
36

KEN
15

Dad's Parents

car accident
at age 32

A?

DAD
59

JON
38

MARY
15

MARY
38

A?

JOE
17

Key

Male

Female

Death

A = Alcohol abuse

A? = Questionable alcohol abuse

what one woman, Libby, discovered.

Libby's Story

"Last week my mother, sister Jean, and I went to visit my grandmother's first cousin, Anna. She's considered the family historian, and I explained that I was trying to learn about our family so I could draw a family tree or genogram.

"Cousin Anna had a great deal to say. I found out a few things I had never known before. For instance, quite a few of the men in our family died very young. Two of my uncles and my great-grandfather died before they were forty years old. They left their wives with families to raise on their own. My widowed aunts and great-grandmother never remarried, and it's no wonder that everyone talks about what 'strong' women they were. Women in our family are usually described as feisty, domineering, and strong-willed. I was always told, 'You come from a long line of strong women,' but I had never appreciated the unusual circumstances that forced most of the women in the family to be so strong and independent.

"Uncle Philip died when he was thirty-nine. He was killed in a tractor accident on the farm. My grandmother's father had a fatal heart attack when my grandmother was twelve years old. My Uncle Victor hanged himself after a big family fight. He was an alcoholic and very abusive. Anna finally revealed this family 'secret.' Until she told us, everyone believed he died as a result of being hit by a car. The truth was he was drunk and my aunt was threatening to walk out on him. That night he hanged himself in the cellar.

"But even with all these horror stories, the family had some good times, too. Anna told us a few wonderful love stories. Each generation seemed to have a tragedy, but each generation also had a Romeo. It was great to learn about family romances. I'm sure these stories were somewhat exaggerated, but they certainly made for great listening. It was nice to think of my relatives as wildly romantic. I had always thought of them as salt of the earth, hardworking types with little interest in matters of the heart."

Libby did a simple genogram, drawing in the names of her relatives in three generations and noting the three most interesting patterns that stood out to her (Diagram 4). These were the themes she became aware of from the discussion about the family that she had with her mother, sister, and grandmother's cousin: premature deaths, strong women, and romantic men.

Libby followed a simple version of the Basic Genogram form, described in the next chapter. She used gender symbols: squares to represent men, circles to represent women. She placed an X through a symbol to indicate death.

Within each figure, she put the person's name and present age. Next to it she wrote the characteristics ("strong woman," "romantic man") that described the personality traits she was tracking. For the "male premature deaths"

DIAGRAM 3—GENDER AND DEATH SYMBOLS

male   female   death

she was noting, she included the age at death of the person within the gender symbol and wrote cause of death next to the square. To indicate her own place in the family, she used a double black line for her circle.

Libby's Findings

Libby's story is a good example of some of the benefits of doing a genogram.

- Discussion with relatives and drawing the genogram brought Libby's family tree to life. For the first time in her experience, Libby remembered more than the names of her deceased relatives. They now had personalities, feelings, and particular life experiences that made Libby feel they were still alive within the history and spirit of the family.

DIAGRAM 4 — STRONG WOMEN AND ROMANTIC MEN

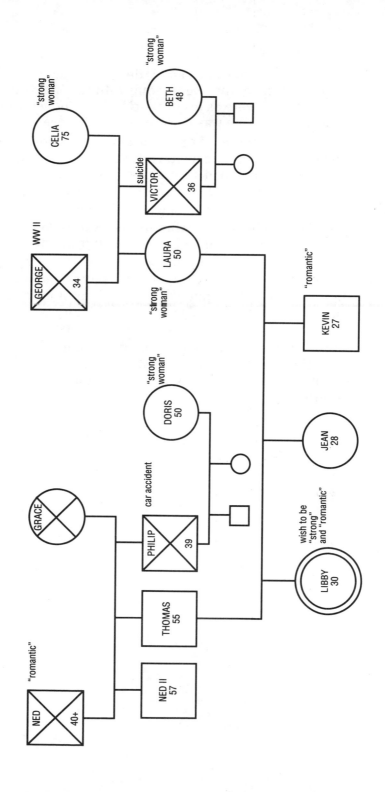

- Any one of the episodes Libby described had an interesting story that went along with the incident. These stories were as exciting to her as a novel, film, or soap opera. She was curious to know more.

- In tracking patterns of premature deaths, strong women, and romantic men, Libby realized that her family, like most, had several themes. In her case, the themes were love, strength, and tragedy.

- Libby felt that she was a part of her family's past. She felt proud to be part of a family that survived terrible tragedies and yet still had a wonderful romantic streak.

- She got to know her living relatives better and was interested in their reminiscences and opinions about the family. Her knowledge of family history was augmented by their contributions and insights.

- Libby said it was fun as well as interesting to put all of these names and events down on paper. For her, it was an entertaining as well as educational experience.

3
THREE TYPES OF GENOGRAMS

What to Include on Your Genogram

You are the master of your genogram, so you get to draw up the master plan. This chapter outlines three types of genograms to use, with certain symbols and keys to follow, but the actual material that you put on the genogram flows from you and your sources of information.

As a basis for selecting a type of genogram, be aware that there are two basic views you can have of your family: an overview and an inner view. The overview or bird's eye view means taking a long view of your family. If you take this view, your genogram will show basic historical facts and particulars:

- Dates of birth, marriage, separation, divorce, death, and cause of death
- Sex and birth order of each family member
- Ethnic background

- Religion
- Occupation
- Education
- Geographic location of living family members

Mary, a thirty-three-year-old teacher, decided to do this kind of genogram for three generations. In less than forty minutes, she completed a simple genogram (Diagram 5).

In this straightforward historical rendering, she came up with the following "coincidences," which illustrate how the bare facts can be full of interesting information.

- In three generations on the maternal side of her family, firstborn children were female (Mary, her mother and grandmother).
- All three of these firstborn women were in the same profession: teaching. Mary taught dance to grade school students, her mother was the director of a daycare center, and her eighty-seven-year-old grandmother had been an elementary school teacher for forty-five years.
- There also were two men in the teaching profession. Mary's brother Kent was a computer trainer, and her Uncle Philip (mother's brother) was a high school teacher.
- Mary's aunt (her mother's sister) was a teacher, too, bringing the teacher total to six.
- Mary's mother and maternal grandmother had both been divorced. Mary is single.
- Mary's maternal grandmother had three children, as did Mary's mother.
- Mary's parents divorced, and both then married partners who were much younger (her father married a woman fourteen years his junior, and her mother married a man eleven years younger).

In contrast to the view Mary took, the microscopic view or inner view gives you a closer look at specific incidents or issues and allows you to see your family in two other

DIAGRAM 5 — MARY'S GENOGRAM

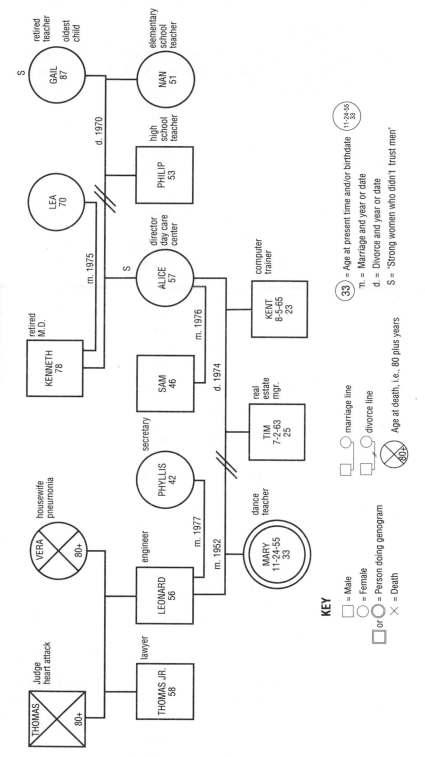

KEY

☐ = Male

○ = Female

☐ or ○⃝ = Person doing genogram

✕ = Death

⊟○ marriage line

⊟☐ divorce line

⊗ Age at death, i.e., 80 plus years

③③ = Age at present time and/or birthdate

m. = Marriage and year or date

d. = Divorce and year or date

S = "Strong women who didn't trust men"

contexts: social and psychological. Genograms with this view include information about a variety of items like emotional and physical problems; family roles; life events; repetitive patterns in functioning, relationships, or behavior; and what happened or is happening in your family now.

Initially, the microscopic view may seem overwhelming. There are virtually hundreds of specifics you can put under the microscope as you set out to draw your family tree. But remember, once you have recorded most of the basics, you can decide to document as few or as many family characteristics as you like. You can keep your genogram very simple. In fact, it's recommended that you do. Too much information is definitely too much to decipher and comprehend.

Bess, a twenty-eight-year-old lawyer, kept her genogram simple by choosing one issue to investigate. She figured out that the issue she was most interested in at the moment was one family theme: money.

She was getting married soon, and she realized that her panic about financial matters was highly emotional and totally unrealistic. Still, she was having a hard time dealing with her fears of not being self-sufficient in all financial matters. She had a vague feeling that most of the people in her family had some kind of "money hang-up." She felt the genogram might give her a better perspective about the issue that was creating so much anxiety for her.

Bess drew a "money genogram" (Diagram 6). It combined basic facts about her family (names, ages, and occupations) with details concerning attitudes about money.

Not surprisingly, Bess discovered that her fears about money came at least partially from the messages she got from her family. Next to each person's symbol on the genogram, she wrote a few words describing that family member's main concern about money. It was not hard to see that she had incorporated some pretty strong money fears from her parents, brother, and grandmother. Seeing several burning money issues on the genogram made Bess realize it wasn't just her "fire" that she was trying to put out. Money problems and issues had been major fears for all the people close to her. She then saw her money prob-

DIAGRAM 6—MONEY GENOGRAM

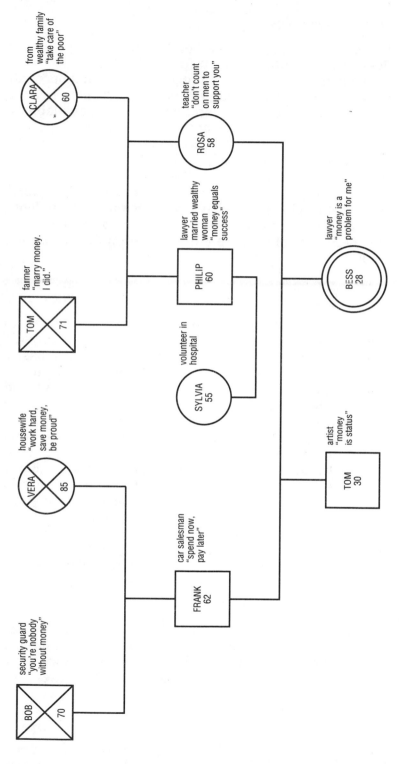

lem in a larger context as a family problem. This helped Bess to stop blaming herself for her fears.

The family genogram outlining the money details made Bess realize that she didn't have to hold on to her old fears forever and that several options were available to her. She could try to let go of the old family message, and she could start to focus on something else every time she slipped back into worrying about money. She made up a new slogan to repeat to herself every time she or anyone in her family started talking about money: "The worst thing that can happen is that I will be temporarily in debt." By drawing a genogram that focused on a particular problem, Bess was able to pinpoint information that suggested a solution.

Three Three-Generational Genograms

Depending on whether you take a broad view or an inner view, your study of your family tree could provide information from one or more of three categories: Basics (names and dates of biological and legal connections), Distances (interactions of people within the family), and Details (character traits, life experiences, themes, and patterns).

Charting all of this information at once could be quite cumbersome. Therefore, you should initially look at your genogram as three separate family trees: A Basic Genogram, a Distances Genogram, and a Details Genogram. You can start by doing a basic genealogical-type family tree, the broad view. Once you've mastered the basics, you can move on to one or both of the other genograms, which contain more specific, detailed descriptions of your relatives and their emotional relationships. The format and symbols you use for each genogram will be the same, but your agenda will color your tree accordingly.

Coming up with your own agenda is the biggest challenge of the genogram exercise. The three genogram types are presented to make your task easier and more organized. They are meant to be suggestions and guides. You might choose only one type of genogram or combine elements from all three.

The Basics Genogram

The Basics Genogram contains the basic facts of your family tree. It describes the vital data about each person in your family, your parents' families, and your grandparents' families. Similar to the traditional genealogical family tree, the Basics Genogram identifies the sex, names, and dates of birth, death, marriage, divorce, and remarriage of all family members. It often shows ethnic backgrounds, noting when foreign-born family members arrived in this country. It specifies religion, occupation, and highest level of education.

It notes biological and legal connections, but it also includes household members who may not have been relatives but were still "part of the family." For one family this was a live-in housekeeper who had been with the family for forty-five years; another family had a "friend who came to visit and never left." Adopted and foster children should be identified, as well as any known miscarriages, abortions, or stillbirths.

The Basics Genogram is the who, how, and when genogram: who was family; how they got to be family (by birth or marriage); when they arrived in the family (birth date, birth order, marriage date, adoption date, placement date, and so on); and when they left the family (date and cause of death, year of separation or divorce).

The Distances Genogram

The Distances Genogram expands on the Basics Genogram by describing interpersonal relationships within the family system. It charts emotional rather than geographical distances. It lets you see how people in the family are connected or disconnected. You find out about the kind of relationships family members have or had by asking yourself and others certain questions:

• Who are you closest to in your family?

- With whom do you have the most conflicted relationship?
- Do you have a close but also conflicted relationship with particular people in your family?
- Are you cut off or estranged from anyone in your family? When and how did this breach occur?

Depending upon whom you ask about relationships and what your own observations are, you will get different perspectives based on different sensibilities. Unlike the Basics Genogram, the Distances Genogram is not based on hard facts but on people's impressions about how relatives behaved and got along in the family. With this type of genogram, you are trying to get a wider picture of the family. It is purely subjective.

When you have drawn in some of these distances on your genogram, you will likely see the formation of patterns in the relationships. The next chapter will discuss ways to interpret these patterns.

The Details Genogram

The Details Genogram diagrams almost any other piece of information you may want to include about family members. This genogram contains specifics about people and events. It can identify medical or physical problems, personality traits, unusual circumstances, or coincidences of life events. Family themes, roles, traditions, beliefs, and feelings can be documented on a Details Genogram. As you can see, many, many items may be of interest to the person doing a genogram.

Don't try to do too much. Don't get embroiled in dozens of details. It is quite sufficient, even preferable, to diagram only one detail on a genogram, explore one particular pattern, or check out one subjective impression. After all, you want to be able to look at your genogram and see patterns and connections.

Your Details Genogram actually becomes a subject or theme genogram. Whether you choose to document particular circumstances or personal characteristics, your

Details Genogram can tell you a great deal—even if it's a one-word adjective about each person in your family.

Now let's look at how to prepare each type of genogram.

Format and Symbols for the Basics Genogram

Like the other two types of genograms, the Basics Genogram is a diagram of vertical and horizontal lines with symbols representing family members. The genogram format and key adapted here follow the standards recommended by the Task Force of the North American Primary Care Research Group as outlined in *Genograms in Family Assessment* by Monica McGoldrick and Randy Gerson.*

DIAGRAM 7

Key Symbols and Lines

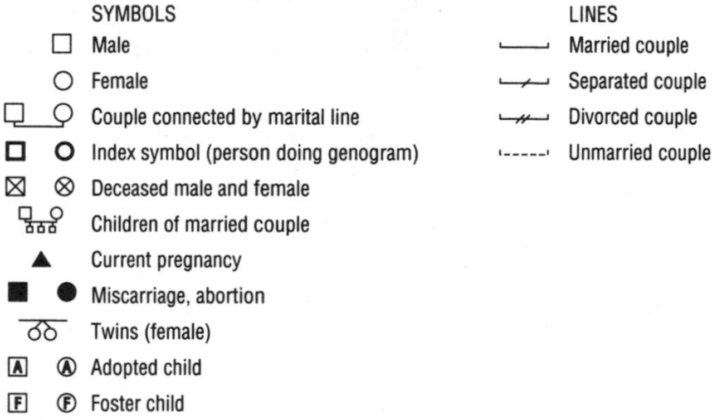

SYMBOLS		LINES	
□	Male	⊢——⊣	Married couple
○	Female	⊢—╱—⊣	Separated couple
⊡ ○	Couple connected by marital line	⊢—╱╱—⊣	Divorced couple
■ ●	Index symbol (person doing genogram)	⊢-----⊣	Unmarried couple
⊠ ⊗	Deceased male and female		
⊡⊤○	Children of married couple		
▲	Current pregnancy		
■ ●	Miscarriage, abortion		
◌◌	Twins (female)		
Ⓐ Ⓐ	Adopted child		
Ⓕ Ⓕ	Foster child		

On the genogram, symbols represent people by sex (circles for females, squares for males, triangles for current pregnancies), and lines show relationships. The basic horizontal line is called a marriage line and connects the male (square) partner on the left with the female (circle) on the right. Vertical lines are drawn below this marriage line to indicate the children from this union. Children are drawn with the appropriate gender symbol in order of birth, from left (eldest) to right (youngest). A three-genera-

---

* New York: W. W. Norton & Co., 1985.

tion genogram can include you, your parents, and grand-parents or you, your parents, and your children. (Sometimes another generation can be loosely sketched in to establish a particular pattern.) A double line is drawn around the symbol of the person drawing the genogram. Thus, if you were the middle child, a daughter, starting to draw your genogram, it would look like this:

DIAGRAM 8—TWO GENERATIONS DIAGRAM

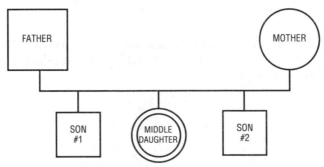

Step-by-Step Construction

Now that you have an idea of the nuclear family format, the following steps outline how to add the information you would include on a Basics Genogram. Assume a family such as the one in the preceding example, and begin by drawing a blank genogram like the figure below.

DIAGRAM 9—BASICS DIAGRAM

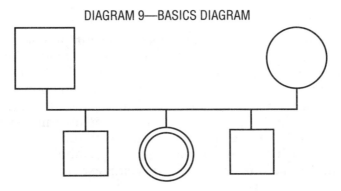

Assuming that the parents are John and Mary Doe, write their names in the square and circle. Then write the date or year of marriage after the letter *m.* on the marriage line, along with the family name (in most cases, the father's last name). If a woman keeps her last name or a couple hyphenates their names, use these names on the marital line.

DIAGRAM 10—MARRIAGE

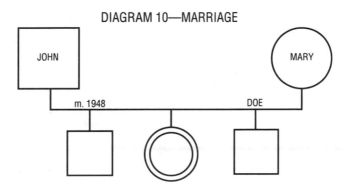

A single slash across the marriage line indicates a separation, and a double slash indicates a divorce. If John and Mary Doe were separated then divorced, the genogram would look like this:

DIAGRAM 11—DIVORCE

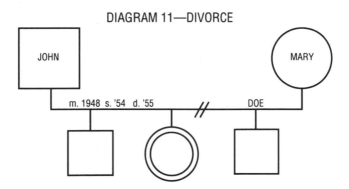

Now the person preparing the genogram is ready to write in her parents' birthdates and the names and birthdates of her siblings. Assume that Marie is the person preparing the genogram, and that her older brother is John Jr. and her younger brother is James. Marie can put the information within or to the right of the appropriate symbol. Within the gender symbol she can put present age and/or birthdate. She can do the same for age at death or date of death. This is how her genogram looks now.

DIAGRAM 12—BIRTH DATES OR AGES

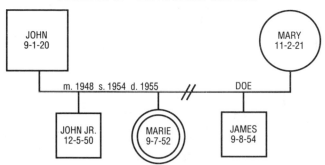

Assume now that Marie's younger brother, James, was killed in a car accident when he was eighteen years old. An X through a gender symbol indicates a death. Wherever possible, write the date and cause of death next to the symbol and put the age the person was when he or she died within the symbol:

DIAGRAM 13—DEATHS

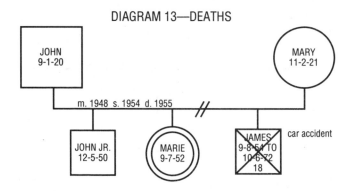

A couple who are engaged or living together but not married, or otherwise established as an unmarried couple, are indicated by a dotted rather than a straight line. Again, the male is on the left and the female on the right. The date the couple met or began living together should be indicated above this line:

Multiple marriages cannot always be drawn in the same format as one marriage. Space is often a problem. First draw the marriage that includes the person drawing the genogram, then any earlier or later marriages. Following is an example of a couple in which each spouse had a previous marriage:

DIAGRAM 15 — REMARRIAGES

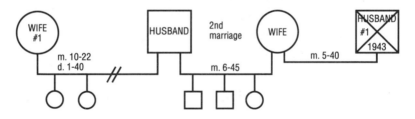

This genogram shows that Husband and Wife were married in 1945 and have two sons and one daughter. Wife was previously married to Husband #1, who died in 1943. They had no children. Husband was previously married to Wife #1 and was divorced from her in 1940. He has two daughters by Wife #1.

Practice completing a Basics Genogram by drawing in the following information on a blank genogram. Start by copying the information added so far to the Doe family diagram. Then, by adding additional symbols and lines and writing notes next to the applicable symbols, incorporate the following data:

— John's (the father's) parents, James Doe and Sarah Smith, married in 1915.

— James was eighty when he died of a stroke.

— Sarah died of a heart attack at age seventy.

— James Jr., the oldest child of James and Sarah Doe, was born October 1, 1916.

— John Doe was the second child, born September 1, 1920.

— Louise, the youngest child of James and Sarah, was born July 2, 1924.

— Mary's (Marie's mother's) parents, Samuel Wilson and Mary Grant, married in 1919.

— Samuel died of cancer at age forty-five.

— Mary Grant Wilson died of a stroke at age seventy-three.

— Mary Wilson Doe's two younger sisters, Ann and Sue, were born June 8, 1923, and June 10, 1925.

— The Doe family is Irish and Roman Catholic.

— The Wilson family is English and Episcopalian.

— James Doe, Sr., and James Doe, Jr., were carpenters.

— John Doe is an engineer with a master's degree.

— Louise Doe is a teacher with a B.A. degree.

— Sarah Smith Doe and Mary Grant Wilson were home-makers.

— Mary Wilson (Marie's mother) was a secretary with two years of college.

— Ann Wilson is a college graduate and teacher.

— Sue Wilson is a high school graduate and a salesper-son.

— John Doe, Jr., is a lawyer.

— Samuel Wilson was a self-employed contractor.

Compare your completed genogram with the following genogram. Does your genogram, for example, show a pattern of people choosing similar occupations?

DIAGRAM 16— COMPLETED BASICS GENOGRAM

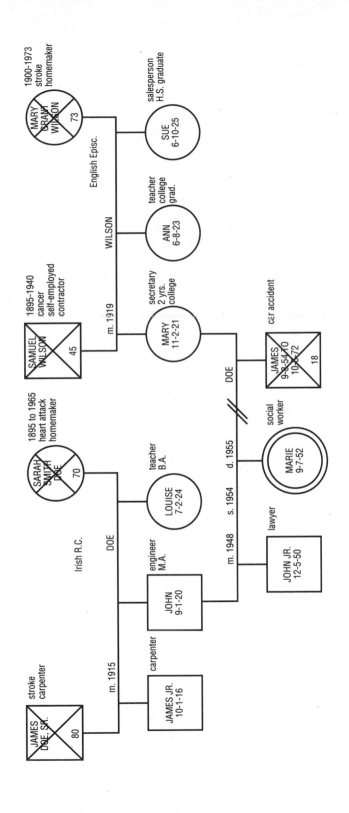

The Distances Genogram Symbols

Once you know how to draw your family tree and put all the people in their proper places in a Basics Genogram, you may become interested in the relationships that exist, or did exist, between people in your family. You can use different lines to illustrate what kinds of relationships human beings have in families. Connection lines between people can depict conflict, closeness, distance, and estrangement.

Drawing these connections creates a Distances Genogram. Here is the key for doing so.

Diagram 17—Key Connections

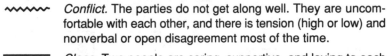 *Conflict.* The parties do not get along well. They are uncomfortable with each other, and there is tension (high or low) and nonverbal or open disagreement most of the time.

======= *Close.* Two people are caring, supportive, and loving to each other.

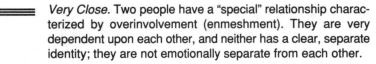 *Very Close.* Two people have a "special" relationship characterized by overinvolvement (enmeshment). They are very dependent upon each other, and neither has a clear, separate identity; they are not emotionally separate from each other.

• • • *Distant.* Emotional distance is evidenced by people who go out of their way to avoid or ignore each other. An indifferent attitude prevails.

——//—— *Cutoff.* A definite breach has occurred, and two people are estranged from each other. There is unresolved emotional attachment denied by separation, withdrawal, running away, isolation, or refuting the still intense connection. There may be no contact, but there is still a very strong tie to a person.

Each person has one of these kinds of relationships with each person in his immediate as well as extended family. These relationships exist in and between generations. Figuring them all out will get complicated, but you can draw in the connections that seem to have the most impact on your life and seem to have been major influences in your family. Most obvious will be the relationships with the most conflict, closeness, or estrangement.

Following is a genogram with relationship lines drawn in, making it a Distances Genogram.

DIAGRAM 18 — RELATIONSHIPS GENOGRAMS

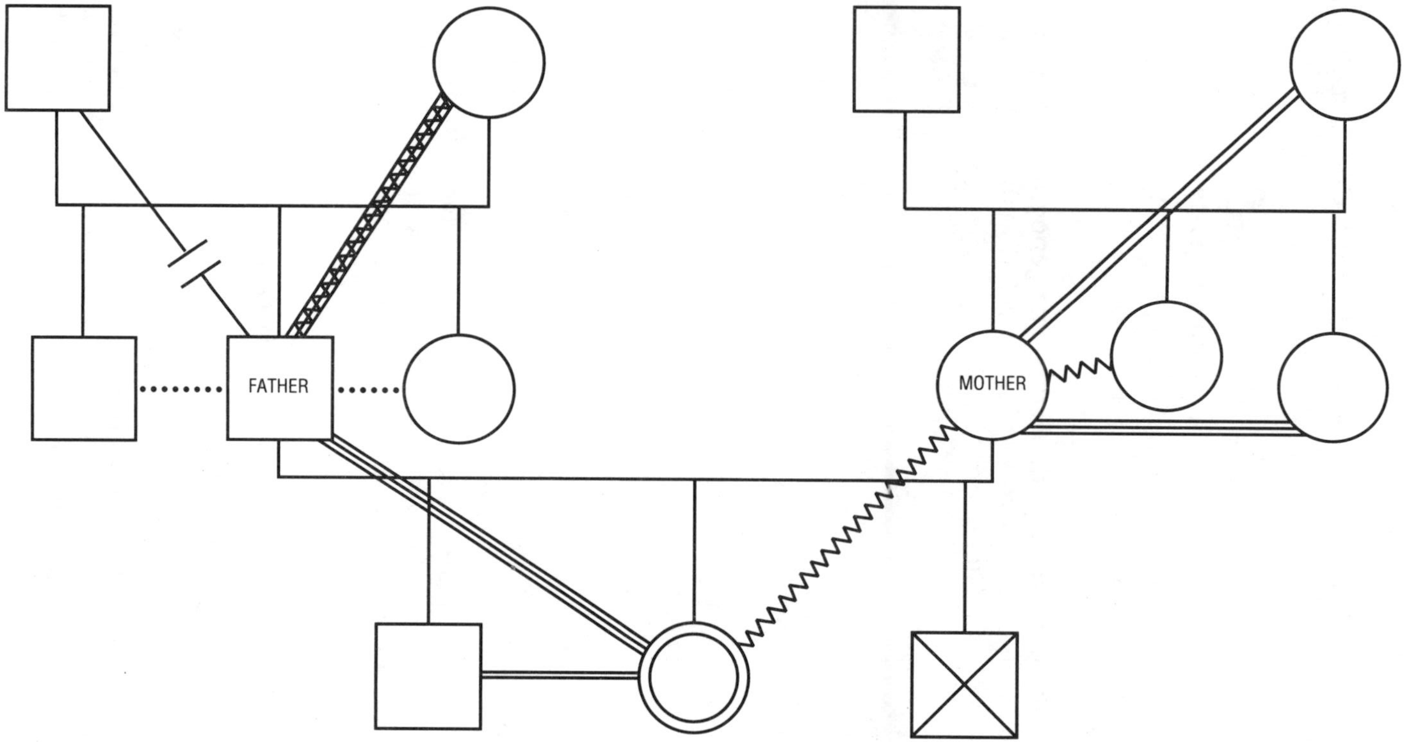

As you can see, Mother has the following notable relationships:

- Close with her mother
- Extremely close with her youngest sister
- Conflicted with her next younger sister
- Conflicted with her daughter

In contrast to Mother's distance genogram which shows two close and two conflicted relationships, Father has mostly conflicted ones:

- Distant with both siblings
- Cutoff with his father
- Extremely close but conflicted with his mother
- Extremely close with his daughter

The person who did this genogram (the middle child, a female) has the following relationships:

- Close with her older brother
- Conflicted with her mother
- Extremely close with her father

Think about the relationships you have with the people in your family. Drawing a Distances Genogram will give you an idea of how close or distant you are from your nearest relatives.

Doing a Details Genogram

Preparing a Basics Genogram will probably prompt you to ask yourself all kinds of questions about your family. Perhaps you have some particular impressions or issues that appear to have a three-generational pattern or theme. Noting facts such as the occupations of people in your family may suggest a pattern. For instance, this could be a a "caretaking tradition," in which relatives have been involved in jobs where they took care of others, such as teaching, nursing, or counseling.

The Details Genogram helps you to check out how events in your life are similar to those that happened in another generation. You may ask yourself whether these are mere coincidences or particular family patterns.

Some details you might begin to think about in terms of your family are:

Affairs	Inconsistencies	Pressures
Alcohol abuse	Indecision	Public service
Anxieties	In-law trouble	Relationships
Atmosphere of	Insensitivity	Religion
family (emotional,	Intermarriage	Remarriage
political, or social)	Intimacy	Rescuing
Attitudes	Intrigue	Resilience
Behavior	Jail	Roles
Birth order	Jealousy	Romance
(oldest, middle,	Labels of identity	Scholarship
youngest, adopted,	Legal problems	Secrets
or foster children)	Manageability of life	Security
Career choices	Marriages	Sensitivity
Catastrophes	Martyrs	Separations
Coincidences	Mediators	Shyness
Conflicts	Medical histories	Siblings
Deaths	and problems	Sickness
Dependency	Migration	Single parents
Depression	Military services	Sociability
Disease	Misunderstandings	Stories
Divorce	Money	Suicide
Drugs	Mysteries	Temperaments
Eating disorders	Myths	Tension
Empathy	Nicknames	Traditions
Entertainment	Occupations	Tragedy
Escapes	Outgoing	Trauma
Fears	personalities	Triangles
Finances	Parent-child	Trouble
Fun	relationships	Trust
Functioning	Perfectionism	Violence
Illness	Politics	Willfulness
Incest	Power	Work

As you read through the list, take note of the items that

arouse the most interest, anxiety, or curiosity. These may become the focus of your genogram.

Two Revealing Examples

Juliet Sampson and Ella North are two women who decided to do Details Genograms, choosing particular issues for investigation. Juliet chose fear as a theme in her family, and Ella chose mediation, since that was her role at work and in her family. This is what each of them learned about themselves and their families when they constructed their Details Genograms.

The Sampson family is an example of a family with a history of fear. Juliet Sampson, a thirty-five-year-old personnel director for a large hospital, discovered some of these fears when she decided to draw her family tree. She started out by doing a combination of Basics and Distances Genogram (Diagram 19).

Some of the facts on her Basics Genogram gave her the impetus to then construct a Details Genogram (Diagram 20), which she called her "fear tree." She began to suspect she was not the only fearful person in her family. She redrew the outline of her Basics Genogram. Then, next to each family member, she wrote what she thought was each person's main fear.

These are some of the memories and realizations that this exercise evoked for Juliet:

"I wasn't the only scared person in the family, but, of course, I wasn't aware of that when I was growing up. My genogram revealed that everyone had fears. I thought I was the most timid soul alive. I was painfully shy and literally clung to my mother's skirts. Of course, I didn't tell anyone about these fears.

"My father was a big man, a very imposing figure, especially when he drank. He was an alcoholic—the functional, nonviolent type. However, he lost that cool reserve and would get very tipsy and bump into things when he was drinking, knocking furniture over, things like that. My mother, sister, and I learned to stay out of his way when he was under the influence of alcohol. He never hurt any of us, but we were still afraid of him. He would playfully

DIAGRAM 19 — JULIET'S GENOGRAM

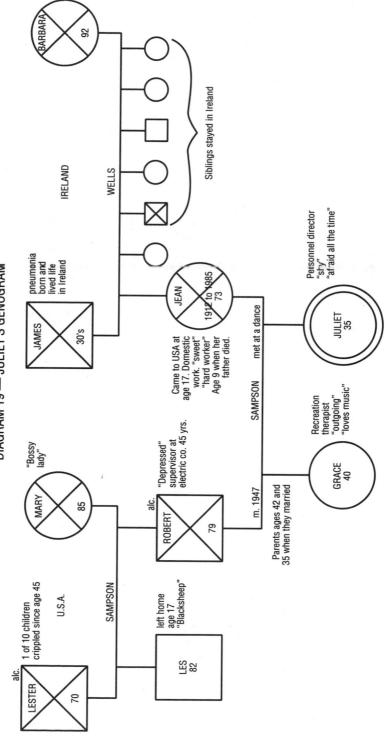

IRELAND

WELLS

BARBARA 92

Siblings stayed in Ireland

pneumonia
born and
lived life
in Ireland

JAMES 30's

JEAN
1912 to 1985
73

Came to USA at
age 17. Domestic
work. "sweet"
"hard worker"
Age 9 when her
father died.

met at a dance

SAMPSON

JULIET 35

Personnel director
"shy"
"afraid all the time"

"Bossy lady"

MARY 85

ROBERT 79

alc.

"Depressed"
supervisor at
electric co. 45 yrs.

m. 1947

Parents ages 42 and
35 when they married

GRACE 40

Recreation
therapist
"outgoing"
"loves music"

alc.

1 of 10 children
crippled since age 45

U.S.A.

SAMPSON

LESTER 70

left home
age 17
"Blacksheep"

LES 82

DIAGRAM 20 — FAMILY FEARS

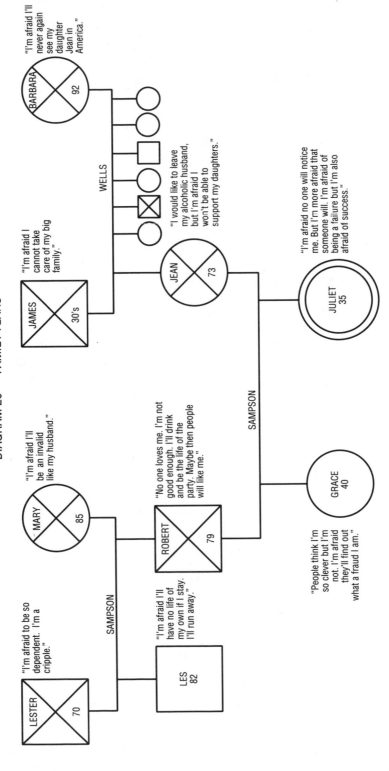

"I'm afraid I'll never again see my daughter Jean in America."

"I'm afraid I cannot take care of my big family."

"I would like to leave my alcoholic husband, but I'm afraid I won't be able to support my daughters."

"I'm afraid no one will notice me. But I'm more afraid that someone will. I'm afraid of being a failure but I'm also afraid of success."

"I'm afraid I'll be an invalid like my husband."

"No one loves me. I'm not good enough. I'll drink and be the life of the party. Maybe then people will like me."

"People think I'm so clever but I'm not. I'm afraid they'll find out what a fraud I am."

"I'm afraid to be so dependent. I'm a cripple."

"I'm afraid I'll have no life of my own if I stay. I'll run away."

WELLS

SAMPSON

SAMPSON

BARBARA 92

JAMES 30's

JEAN 73

JULIET 35

MARY 85

ROBERT 79

GRACE 40

LESTER 70

LES 82

pick me up and hold me in his arms when I was a very small child, and I was sure he was going to drop me.

"I have a sister, Grace, who is five years older than me, and who never escaped my father's anger and verbal abuse. He treated her terribly. I was clearly my father's favorite, which was a blessing and a curse. He seemed to get comfort and security from me, but I never felt safe with him.

"My father's humor was very sarcastic, and I always wanted to protect my mother and sister from his barbs, since they were targets of his verbal abuse. He was particularly nasty to my sister. I felt guilty for being his favorite and tried hard to make up to my sister for all my father's injustices.

"All the attention I got as a child from my father was at the expense of my mother and sister not getting it. It was very painful to be in the position of trying to save them from my father's anger. In his eyes I could do no wrong, and my mother and sister could do no right.

"Comparing the genogram of the past with the present, I see I am still a victim of those fears. In my personal life and at work, I am so afraid of not pleasing people that I go overboard to help and protect them. I'm the person who ends up putting in the overtime and listening to everyone's troubles. I've done well at work and have been promoted for my superresponsibility, but I see that much of my overdoing is based on my fears of disapproval. I was the peacemaker in my family, and I play the same role on my job.

"I'm thirty-five now and have never been married. I have had several long-term affairs with married men. I guess I choose men like my father, men I can never feel totally secure with. And, maybe because I felt so guilty about being favored over my sister, I am always in the other-woman position of being number two. I have never been comfortable with attention. I feel that I want it, but when I finally get it, I feel I don't deserve it.

"My mother had a hard life. She came to this country from Ireland when she was seventeen years old. She worked as an upstairs maid and later as a cook for very wealthy families. She was the oldest of seven children and sent all her earnings back to her family in Ireland. Her father had died in his thirties when my mother was nine years old. All her siblings stayed in Ireland except for one sister who came here and was very close to my mother.

"She didn't marry my father until she was thirty-five and he was forty-two. After my father's death, I learned that he had been furious at my mother for getting pregnant with my sister and did not speak to her throughout her entire pregnancy. He had been the sole support of both his parents, so I guess he was not looking for any more responsibility.

"His older brother left home at an early age, leaving my dad with all the responsibility for his elderly parents. Dad got a draft deferment in World War II because he had to care for them. He seemed to identify my sister with his brother and me with him. Grace was the 'bad child' like his brother, and I was the 'good child' like him."

Ella North's motivation for constructing her genogram was to see why she is dreadfully afraid of being perceived of as a "bad" or "mean" person. She also wanted to understand why she is extremely focused on keeping peace and protecting people for whom she feels responsible.

Ella is an attractive, successful trial attorney in a large, highly respected law firm in Chicago. She graduated from law school with honors, served a clerkship with a federal judge, and is now being considered for a partnership in her firm.

She has been married for ten years, and her husband is loving and supportive. They consider themselves a happy couple. A good marriage, good job, and good prospects have not, however, made Ella feel particularly good about herself. In spite of her high level of competency, she wakes up every morning with a huge knot in her stomach, and her anxiety increases as she goes through her busy day.

Ella is now questioning her choice of careers. Her work is exciting and challenging, and she loves it. But she feels like an imposter. She is fearful that people will find out she is not a "nice" person and that someone will get angry at her. Even though her work puts her in an adversarial position where she cannot please all of the people all of the time, this is what she longs to do. To understand herself better, she constructed a genogram (Diagram 21) that identifies the roles and behavior styles of family members.

Ella's peacemaker role during her childhood and her career choice in the justice system seem congruent. The past explains the present. Does your family history explain any of your actions and aspirations? You might ask yourself whether you identify with any of the family dynamics of the North family. In Ella's case, her family background influenced the way she felt about her job. Considering your family background, you may find similar links between family relationships and your dealings at work.

People are more alike than they are different; so are families. By comparing the Sampson family with the North family, you can see some similarities in the cast of characters, including:

- A favorite child or "good" child
- A "bad" child
- Mediators and peacemakers
- Alcoholism
- Children more successful than parents
- Mothers who lost their fathers early in life
- Young men who died or deserted family
- Men who took care of their parents

In what ways is your family like the Sampsons and Norths? Do people play similar roles or make career choices based on particular family dynamics?

Repetitive Patterns

Not only might you uncover a variety of subjects on a Details Genogram, you may also find that these

DIAGRAM 21 — PEACEMAKER

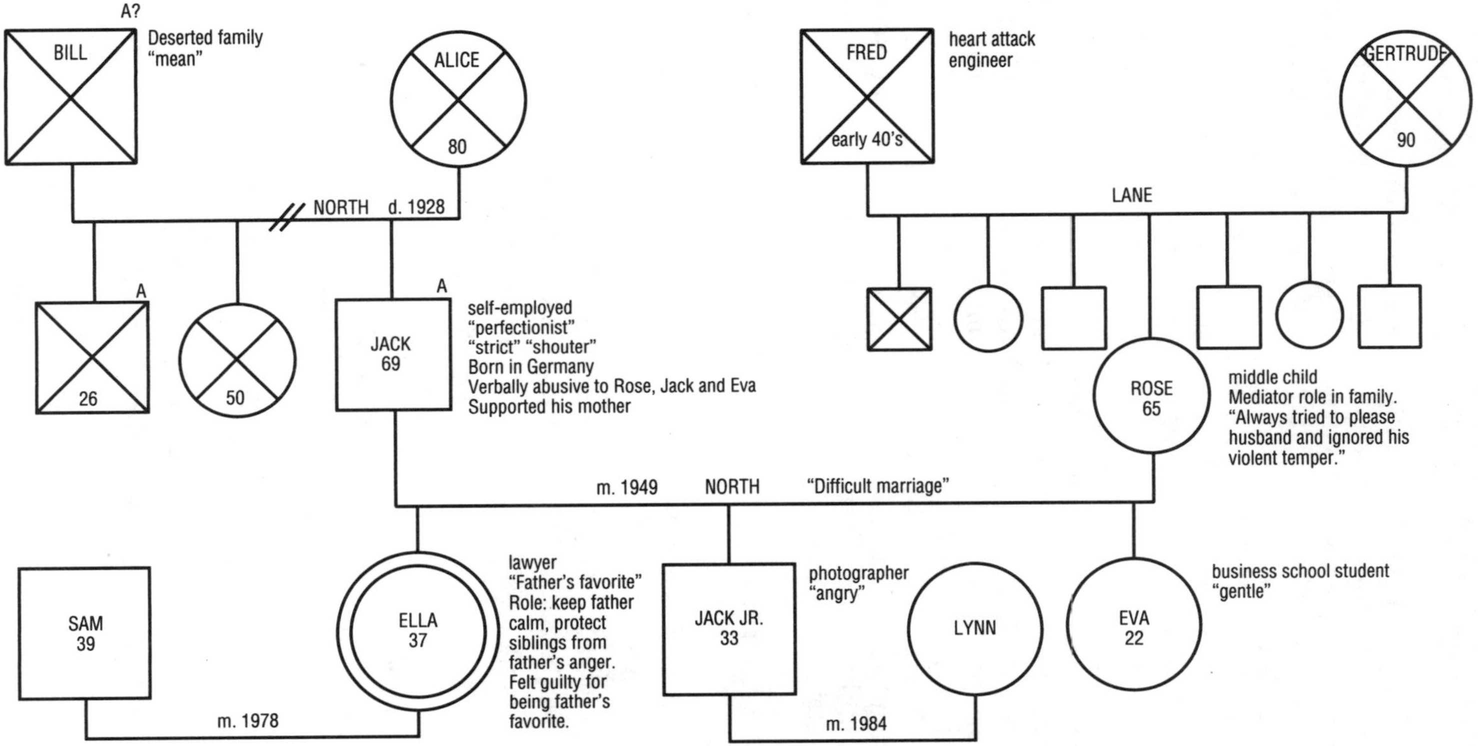

genograms make repetitive patterns graphically evident. For example, LeeAnn, a twenty-eight-year-old schoolteacher, became curious about the prevalence of certain occupations in her history and the occurrence of the disease of alcoholism in several generations. Her Basics Genogram (Diagram 22) clearly showed a pattern of choosing teaching or engineering as occupations, and she decided to enlarge it to see if there was also a pattern of alcohol abuse. Look at LeeAnn's genogram to see how these repetitive patterns show up.

LeeAnn's conclusions were revealing and enlightening. She comments, "It was very interesting to me to find out how women in the family tended to become schoolteachers, while men chose careers in engineering. Three generations produced three schoolteachers and two engineers.

"To my surprise, I discovered that there were several other patterns of coincidences: My father's mother and my mother had the same first name (Elizabeth), and both of these women lost their mothers when they were about two years old. My great-grandmother on my father's side died in childbirth, and my grandmother on my mother's side died of tuberculosis. Also, both of my parents were only children. Although my father had a stepsister, she was fifteen years younger than he, so basically he was raised as an only child.

"Yet another pattern was that my maternal grandfather and my paternal grandmother had each married three times. This is quite an extraordinary coincidence for that time and place in history.

"As I suspected, alcohol abuse showed up often. There are at least six problem drinkers that I found out about: my great-grandfather, grandfather, father, and stepbrother on my father's side of the family, and my maternal grandfather. Also, my father's first wife married an alcoholic after she divorced my father."

In this way, the Details Genogram allows you to find out about particular family traits and patterns over the years. It becomes an analytical tool in addition to being a genealogy. Your current problems or dilemmas may begin to make sense within the context of your history. In con-

DIAGRAM 22 — TEACHERS AND ENGINEERS

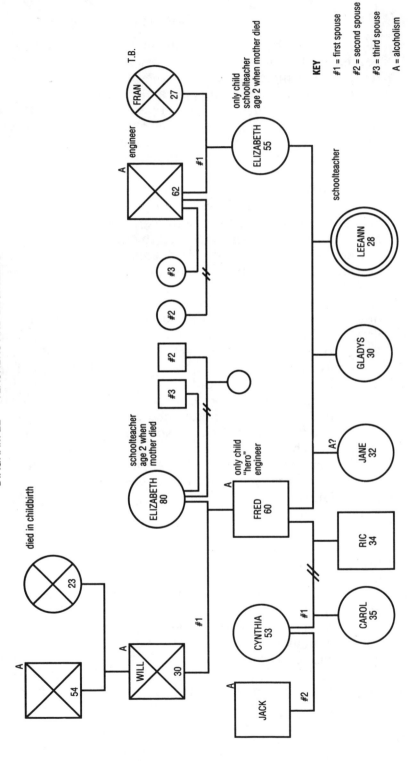

died in childbirth

engineer

T.B.

A

only child
schoolteacher
age 2 when mother died

schoolteacher

schoolteacher
age 2 when mother died

only child
"hero"
engineer

KEY

#1 = first spouse

#2 = second spouse

#3 = third spouse

A = alcoholism

structing a Details Genogram, you have the opportunity to explore your history: special events, personal characteristics, family problems, family themes or traditions, or whatever information is of special interest to you.

It is helpful to keep a log alongside the genogram form and write down the chronology of family events. Listing important events in chronological order can help to explain certain individual behaviors or other events. For instance, one thing that is not unusual in families is a marriage date the year of or year after the death of a significant relative.

Both your genogram and family log can be as simple or as complex as you desire. The amount of historical information available to you will also influence the scope of the graphic and chronological listings you undertake.

Extended Family Genograms

The three-generation genogram is recommended because it is a manageable type of genogram, which most people can do. As you know by now, genograms can get complicated very fast, particularly if there are many divorces and remarriages. Also, many people do not have the requisite information to complete genograms beyond three generations. However, there is nothing particularly magical about the number 3.

Perhaps you really want to expand your genogram beyond three generations. By all means, go ahead. The more you seek, the more you will find. After doing three generations, it's not at all unusual for people to want to find out more about their ancestry.

Thomas, for example, is a Civil War buff. He has always been fascinated by that period in history. But before doing his own genogram, he had never really thought much about the Civil War in relation to his family. Noting that his father and grandfather had military backgrounds, he was curious about how far back this military service went in his family.

To his surprise, Thomas found that he came from a long line of soldiers on both sides of his family. So far he has counted fifteen soldiers in five wars on his genogram.

Soldiering is becoming more relevant in his family genogram which documents almost five generations.

Some people have influences outside the family that are important in their personal histories. It is often illuminating to include "other families" in your family research. You can attach these genograms to your own to compare such influences, thus expanding your three-generational genogram in another direction.

Celia is one person who did this. She is an only child of a dual-career couple. Both parents worked in demanding jobs, but her mother had the more successful career. She worked long hours while Celia was growing up. Consequently, Celia spent a great deal of time with the next-door neighbors, the Jones family. They were, in her words, her "surrogate family," the family she had always wanted. She felt that the five Jones children were the brothers and sister she never had. Celia was at the Joneses' house every day after school, frequently dining with them on her parents' late nights.

Because of the strong influence of the Jones family, Celia extended her genogram (Diagram 23) to include this family who meant so much to her growth and development from grade school through high school. She drew the Jones genogram over two generations and stapled this to her own. It became clear that there were indeed many characteristics in her makeup that could easily be attributed to her other family. From the Jones family she developed an interest in sports and athletic abilities. She learned about competition and team play, which later helped her in her career.

There were four boys in the Jones family, all a few years older than Celia. Mr. Jones and his sons were baseball fanatics. They supported the local farm team of a major-league club and went to all the home games. Celia and Betsy Jones (the youngest of the Jones children and Celia's best friend) went along, too. The girls also got to play ball when the boys and their cousins and friends chose up sides. In typical sexist fashion, the girls were last chosen and mostly assigned left field, but they did get to play, and they enjoyed themselves. "I guess it's no surprise that I'm captain of our company softball team," says Celia.

DIAGRAM 23 — SURROGATE FAMILIES

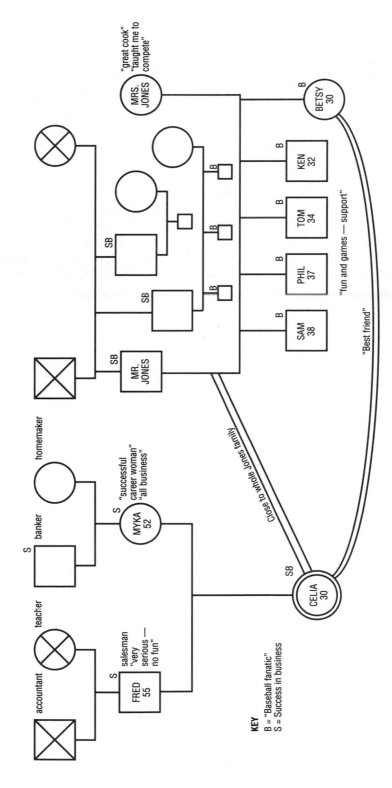

KEY
B = "Baseball fanatic"
S = "Success in business"

Celia loves to cook, and she claims that this is a Jones family influence also. "Mrs. Jones made the best lasagna in town," remembers Celia, and notes that her mother was a "lousy cook." Mrs. Jones taught Celia and Betsy how to cook and encouraged them to submit their cakes in 4H competitions. Celia won several ribbons for her chocolate cake.

Before she completed her genogram, Celia had thought she took after her mother, since she strongly identifies with her mother's successful business career. But adding the Jones family genogram increased her perspective. She saw how the influences of team play, competition, and winning and losing had all added to her abilities and self-confidence. Her business savvy is from her real family *and* her surrogate family.

4
TRIANGLES ON YOUR GENOGRAM

Troubled Threesomes

When you draw relationship lines in a Distances Genogram, triangles will probably leap off the page. Drawing the relationship lines helps you to see how threesomes of people form the basic building block of all emotional systems. Trouble usually becomes obvious when viewed within the structure of a triangle, particularly in your own family.

Despite the importance of triangles, our romantic souls hint that we go through life in twosomes. Our language and views reinforce this. We tend to view ourselves as parts of pairs or couples, and often think of two-party systems like mother/child, husband/wife, and father/son. Yet as much as we would like to be viewed in twosomes, we are just as often—actually more often—in triad rather than diad situations.

The term *triangle* in most people's vocabularies describes a relationship with an illicit connotation. When most people hear the word *triangle*, they think of the "eternal triangle" made up of a married couple and the third person with whom one partner is having an affair. We are not used to thinking in terms of other kinds of triangles, but there are many triangles within family systems.

As you draw your genogram, you start from a couples point of view. You begin with the couple that produced your family—your parents. But each partner in every couple comes from another family. The twosome is never just a twosome.

Consider an example of a common triangular situation that exists in almost every family, with only slight variations in theme and process. This hypothetical case illustrates how triangles are formed and how they shift and grow as situations increase from low to high stress. Mother and father are the parents of three grade-school-age sons, Adam, Billy, and Chuck. Mother and Father have been called into the principal's office to discuss the behavior of their middle child, Billy. They try to figure out how best to handle Billy and his behavior problem. We can draw this relationship as a triangle (Diagram 24).

DIAGRAM 24

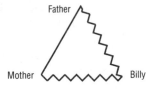

After this talk, father goes off on a business trip, and mother has sole responsibility for disciplining Billy, who begins to have behavior problems at home. He is constantly fighting with his younger brother, Chuck, and mother tries to referee (Diagram 25).

Billy and Chuck are in constant battle, so Adam, being the eldest, joins in to try to protect baby brother Chuck. All three boys are now in a very competitive situation (Diagram 26).

When father returns home, he is at a loss as to how to handle this increasingly volatile situation, so he tele-

DIAGRAM 25

DIAGRAM 26

phones grandfather, a retired schoolteacher, to enlist his
help and ask him to speak with Billy about this unaccept-
able behavior (Diagram 27).

DIAGRAM 27

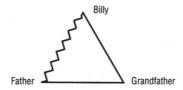

Drawing these triangles on the genogram will look like
Diagram 28.

As you can see, there are four distinct triangles in the
Jones family around just one problem.

Different problems will create different triangles. Various combinations of any close and conflicted relationships produce eight possible types (Diagram 29).

And, as shown in the Jones family, a member of the
original threesome often finds a fourth person to bring in.
This takes one of the original participants outside the triangle, at least temporarily (Diagram 30).

Triangles detract attention away from a real conflict

DIAGRAM 28 — GENOGRAM WITH TRIANGLES

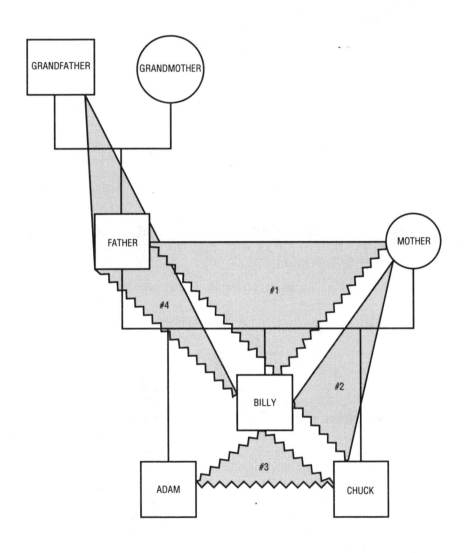

DIAGRAM 29

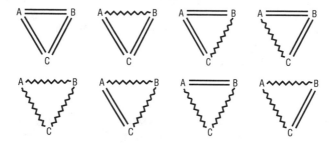

between two people. If one against one feels too dangerous to face, a third person can dilute that potentially
hazardous situation by becoming an additional focus.
Thus, the original knot becomes more tangled and less
recognizable. It is soon lost among the many tangles that
continue to form as the triangles increase.

DIAGRAM 30

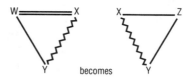

becomes

Triangles rarely solve problems; they submerge them
by creating different problems. Triangles can become
rigid systems that block the resolution of whatever conflict exists between two people. Because triangles help
family members ignore unspoken emotional issues, basic
emotional needs are not met, and broken connections are
not mended. In addition to preventing the resolution of
conflicts, triangles prevent people from growing and becoming the best they can be. If we are always focused on
one or both other parties in the triangle, we do not focus
sufficiently on ourselves. When we are worrying about the
conflict between two others, we do not have a chance to
work on our own particular needs and interest.

For that reason, triangles are a good place to start in
identifying the true source of conflict. Family genograms

show that triangles in one generation tend to resemble triangles in previous generations. Family patterns, in the form of triangles, repeat themselves.

Another reason to look for triangles is that you can change the one you're involved in. Triangles are seldom static. Just as there is a way into triangles, there is a way out.

"Always on the Outside" Triangles

One commonly repeated pattern is the "always on the outside" triangle. In this pattern, a family member is never closely linked to the others in a triangle.

This pattern clearly applies to Sam, a forty-seven-year-old business executive who is going through his second divorce. Interestingly, his reasons for the failures of both marriages are basically the same. Describing his first wife, he says, "She was more interested in her career than in me." About his second wife: "I'm a stranger in my own house; my wife dotes on her kids [Sam's stepsons] and never has time for me." In both cases, Sam feels he doesn't count; career or children are more important to the women he marries, and he feels on the outside.

Is it sheer coincidence that Sam finds himself once again on the outside of a triangular situation? When Sam constructed a Distances Genogram he found out these were not the only situations in his life in which he felt he was left out. In his family of origin, Sam was the eldest of six children and the only boy. This situation created an automatic triangle, with parents, sisters, and himself as the three corners, even before the arrival of any tension or conflict at all in the family (Diagram 31).

Of course, there was tension. Sam grew up feeling isolated and alone. His mother was preoccupied with all the babies in the family, and his father worked the four-to-midnight shift at the local steel mill and was not home much. Even when he was at home, he was sleeping, or reading the newspaper. Consequently, Sam received very little attention from either parent; in that triangle, too, he felt like the outsider.

DIAGRAM 31

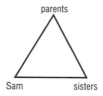

There was one instance, however, in which Sam felt on the inside of a triangle. He was very close to his sister Janice, ten years his junior. He was her special baby-sitter as a child, and as they got older they became even closer. They saw, and still see, themselves as different from others in the family. Sam remarks, "We feel the same about the family, and neither one of us has much contact with any of the others." Sam was able to get himself, finally, to an inside position on a triangle (Diagram 32).

DIAGRAM 32

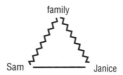

Diagramming the triangles in his family of origin and both marriages (Diagram 33) helped Sam to see yet another triangle. This was in a work situation. Sam was very angry at his firm because he did not receive a Christmas bonus. To make matters worse, one of Sam's partners seemed to be very negative about a pet project of Sam's. Sam felt alienated from his firm and his colleague. He was seriously thinking about changing jobs. A frank discussion with the people he worked with got Sam out of these triangles. Sam learned that no one in the firm had received a bonus because of financial reversals in the company. He was not, as he had thought, the lone victim. His colleague had been distracted by a health problem and did not disapprove of Sam's project at all. He was simply less than enthusiastic because of his personal concerns. Fortunately, the work problems were resolved by open

DIAGRAM 33 — FAMILY AND WORK TRIANGLES

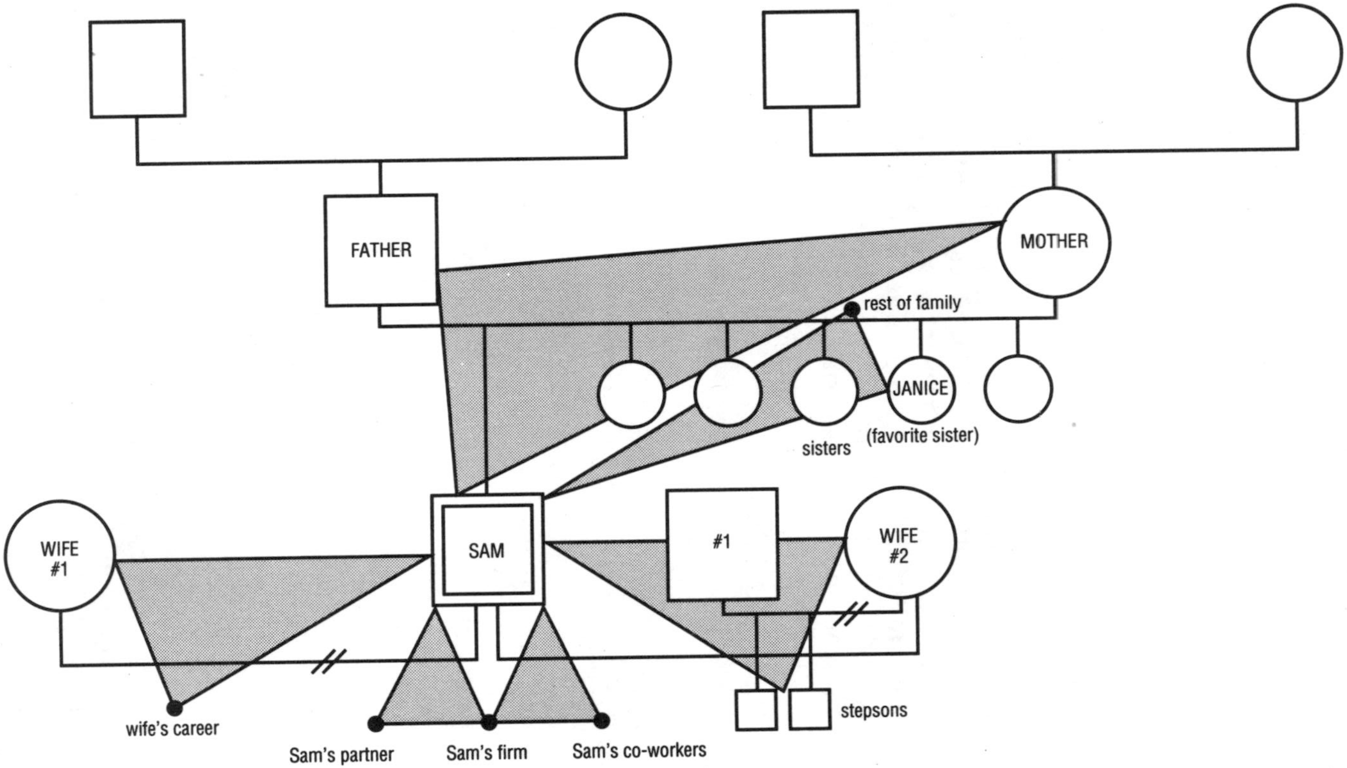

communication, and Sam began to learn how he unconsciously often constructs triangles that put him in the left-out position.

As you can see from Sam's genogram, some of the "members" of his triangles are not people but institutions, jobs, or groups of people. In drawing in the triangles on your own genogram, consider those influences as well. You can add them in the same way on Sam's genogram.

Creation of Triangles

According to family therapist Murray Bowen, M.D., the triangle is "the smallest stable relationship system. A two-party system may be stable as long as it is calm, but when anxiety increases, it immediately involves the most vulnerable other person to become a triangle. When tension in the triangle is too great for the threesome, it involves others to become a series of interlocking triangles."

No family is always calm. Since tension creates triangles, it is easy to see how triangles occur in families.

Triangles often surface when there are power struggles—a common occurrence in any family. Families are not particularly known for their egalitarian practices. Let's look at a few typical power plays and how these create triangles.

If a father is perceived as having all the authority in the family, he's "the boss," and his wife and son may collude, perhaps unconsciously, in an attempt to usurp his power. They may join together in order to have a voice or a stronger voice in the family. The real problem may be a marital conflict between Mom and Dad over some other issue, but by using the son to gain more power, the mother focuses on the child and off the marital problems (Diagram 34).

DIAGRAM 34

son

mom ⌁⌁⌁⌁⌁ dad

Couples often have unidentified marital issues that they are afraid to address for fear the marriage will be in trouble. The original conflict may not even be conscious. Husband and wife cannot face each other, so they bring in another person, often a mother-in-law. She is then identified as "the trouble," so the couple are protected from looking at the *real* trouble, their relationship (Diagram 35).

DIAGRAM 35

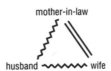

Almost anything or anybody can be drawn into the emotional two-party system to form a triangle. Sometimes issues rather than people become the third side of the triangle. Addictions like alcoholism become the primary focus between two people. The disease becomes the recipient for all the anger and anxiety. Years go by with the addiction getting all the attention, and the couple ends up with no idea about their own issues in the marital relationship.

Useful and Harmful Triangles

Not all triangles cause problems; some are less harmful than others. Sometimes they are temporary, expedient, and cause no trouble at all. We can even laugh in retrospect about triangles we create to misdirect our minor annoyances. For instance, Jane acts furious at her bank for its insufficient services when she is really furious at her husband, Joe, for playing poker with the boys and losing the rent money. It's a lot safer to get angry at First National than at Joe, who might bring up Jane's compulsive shopping (Diagram 36).

Triangles can, however, cause big problems. They can easily become painful and repetitive. If Jane continues to

DIAGRAM 36

blame a financial institution rather than discuss how they go about paying their bills, the couple's money problem will get bigger. Avoidance may work some of the time, but not all of the time. If we do not identify our real problems, we cannot solve them.

Identifying Personal Triangles

Drawing triangles on your genogram may be the first step in stopping old, predictable behaviors. You may gain clarity about the triangles in your life and in the lives of those who were your role models. What you saw while you were growing up was what you learned. What you see now, on your genogram, can begin a process of unlearning or relearning.

As you draw your Distances Genogram, consider the following eight points. They will help you identify your personal triangles.

1. Wherever you have a threesome in which the functioning of two people influences the functioning of a third, you have a triangle.
2. When two people are joined in some kind of battle with a third, you have a triangle.
3. When there is a shifting three-person relationship with one person always in the "out" position, you have a triangle.
4. If you have a very strong emotional reaction to one person and turn for solace and support to another, you may be in a triangle. Strong feelings, particularly negative ones (like anger), are often a clue about triangles. And "off the wall" emotions that seem out of

proportion to current situations may have to do with old triangles—that is, unfinished business in your family of origin. If the anger, say, at a boyfriend, seems way out of proportion to what he is doing, his behavior may be reminding you of similar situations long past. What may at first glance seem like today's triangle (you, your boyfriend, his mother) may not be nearly as provocative as yesterday's triangle (you, your brother, your father). You are angry at your boyfriend because you see his mother spoiling him, but you are really still furious at your brother because your father ignored you and doted on him.

5. If you are overly involved in trying to solve one person's problem with someone else, you are probably in a triangle. You may have been included because the other members don't want to or know how to resolve the conflict between them. Feeling manipulated is another clue that other people are using you in their power struggles.

6. If words like *mediator, fixer, rescuer,* or *buffer* seem to describe your role, chances are you are part of a threesome.

7. If you feel stuck in a relationship or cannot seem to let go of certain feelings, like anger or resentment, it may mean that you have formed a triangle to avoid confronting these negative feelings. Feeling stuck with your fury at a business partner may mean you are really not confronting your anger at your love partner. It may feel safer for you to explode at the third person.

8. Triangles seem to occur more frequently when we are going through some kind of major life stress. Our normal life cycle contains certain acknowledged pressure points when we should be tuned in to the possibility of triangles. These points include: courtship and marriage, birth of a first child, children starting school, children leaving home, a job change or retirement, aging parents, and death.

Escaping Harmful Triangles

It is probably much easier to know you're in a triangle than to get out of one. Triangles are complex, and so are the reasons why they are so immovable. The following list includes reasons why people have a difficult time extricating themselves from triangular situations:

- Fear of separation
- Fear of abandonment
- Fear of conflict
- Fear of change
- Guilt feelings
- Fear of rejection or retaliation
- Fear of anger
- Fear of the unknown
- Fear of confrontation
- Fear of self-examination
- Need to protect oneself and others

Despite these fears, we can benefit greatly from working our way out of most triangles. In her book, *The Dance of Anger,** Harriet Goldhor Lerner, Ph.D., cautions, "All of us are vulnerable to intense, non-productive angry reactions in our current relationships if we do not deal openly and directly with emotional issues from our first family, in particular, losses and cutoffs. If we do not observe and understand how our triangles operate, our anger can keep us stuck in the past, rather than serving as an incentive and guide to more productive relationship patterns in the future."

Certainly, this is an excellent reason for plotting triangles on a family tree. Other benefits include:

- A golden opportunity to deal with "real" issues
- Freedom from negative emotional energy

---

* New York: Harper & Row, 1986, p. 180.

- Acceptance of responsibility for oneself and relinquishing superresponsibility for others
- Emotional relief to be out of a "middle" position
- A chance for true intimacy as understanding grows and old, destructive triangles are dissolved
- Release from negative feelings and unproductive overinvolved situations
- Discontinuation of a negative family pattern by solving problems rather than disguising them
- Creation of a healthy role model for others in the family

If you wish to vacate a particular triangle, try following these four steps:

1. *Decide*—Be aware that you are in a triangle and that it is best for you to be out of it. Your anxiety level is your best gauge to tell you when you need to disengage. The higher it is, the more important it is for you to find a solution. Make a conscious decision to get out of the triangle.

2. *Defuse*—Look at the amount of emotional energy you're expending in a troubled triangle. If you are constantly reacting to what the other two parties in the triangle are saying and doing, you're depleting your energy needlessly. Calm down and cool off. Try to stop reacting. Be a listener and observer rather than an emotionally involved participant.

3. *Detach*—Stop trying to control other people. Realize that you really don't need to fix anything or anybody. Take your focus off the other two sides of the triangle, and put the focus on yourself. Affirm your own importance and autonomy. Consider your own emotional needs and opportunities for personal growth.

4. *Distance*—You may have to temporarily physically remove yourself from the two parties in order to get the emotional distance you need to separate. You aren't fleeing forever; you're just changing the constellation. Be prepared for hard pulls to get you back

inside the triangle. Get all the support you can from outside, disinterested parties.

Getting out of a particular triangle doesn't mean you are cutting yourself off from whatever situation you happen to be in. You are, in a sense, disconnecting (by deciding, defusing, detaching and distancing) so that you can reconnect in a healthier, happier way.

Ben Jones is the adult son of parents who are in the process of a divorce after a thirty-five-year marriage. Since he's the only son and a medical doctor to boot, both his parents look to him to "fix" everything from their marriage and their personal emotional pain. Ben's phone was ringing off the hook with distress calls from both his parents. Clearly, he was in a troubled triangle.

Ben *decided* he was not going to take sides and he was going to get out of the middle. The first thing he did to *defuse* his own emotional involvement was to stop spending two hours a night on the phone with them. He left his answering machine on with the message that he would return the calls at a certain time. When he made his call backs, he told each parent he had fifteen minutes to talk and then would have to end the conversation because he had a "pressing personal matter" to attend to. He didn't have to tell them the "matter" was a long, hot shower he needed to relax himself.

Ben *detached* by limiting the time and energy he would devote to his parents' predicament. Since he could not fix them or their marriage, he referred them to a well-qualified marriage therapist. Whenever they tried to get his advice, he would say, "this sounds like something you should discuss with your marriage counselor..."

Ben's *distancing* was not physical as much as emotional. But when he saw them indivdually he made it quite clear that he did not want to hear about the other parent. Since it was difficult for his folks not to talk about the breakup, he planned his time with them to include activities like movies and shows so he could be with them with a minimum of talk-time. Sometimes he invited his friend Jim along and that also helped to diffuse the loaded emotional situation. Ben found ways to "be there" without being "all there." He needed his distance and he needed to detri-angle.

5
FAMILY TRADITIONS

Folklore and Family Lore

Stories passed down from generation to generation make up your family's special folklore. These stories you cherished since childhood have remarkable staying power. They depict strong personalities, pertinent events, and important facts or illusions that your family has considered worthwhile to preserve.

You can't always be certain whether these stories are fact or fantasy. Most likely, they are a combination of both. The real importance of these folktales is not so much their authenticity but their role in giving your family and you an identity. They reveal what people in your family have felt warrants repetition. Whether these stories are funny or serious indicates whether your family has a sense of humor or a sense of foreboding. They also tell whether your family was concerned with triumph or tragedy or whether, for instance, glory or fear was a common family theme.

Family folklore makes families feel unique. It heralds special traditions or themes that imply a particular family flavor. These tales, repeated over and over again, impart a certain set of values, beliefs, and traditions that usually promote a feeling of kinship. We feel connected. And, generally, we are affirmed and validated by our places in our families and in society.

When these stories create the same theme or same mood, they reveal important family traditions. If all your family stories are about personal triumphs over horrendous troubles, you might surmise that your family values perseverance and success. If the stories are jokes on or about people, humor is probably an important ingredient in your family's character. Being able to laugh at oneself is a big strength in your family. If most of the stories are about the trials and tribulations of unrequited or true love, you can surmise that a romantic streak runs through your family.

Think about all the family tales you remember from childhood to now. Sometimes family stories are about how people got their names or nicknames, how couples met and fell in love, or how people survived terrible blizzards or horrific wars. Write down the stories as you remember them. Observe certain repetitive themes. Compare these themes with the events on your genogram. If, for instance, your family stories tell tales of high adventure and serious risk taking, you might think about the people in your immediate and extended families. Identify them on your genogram as "risk takers" or "risk fearers," writing either description next to the symbols for these people. Is the spirit of your ancestors evident in the lives of your living relatives today? And how does this folklore reflect your personal attitudes and activities?

Traditions and Patterns

Folklore usually connotes tradition. The family stories you heard as a youngster may have carried messages of family expectations—codes of conduct passed down from generation to generation. Family folklore often prescribes and proscribes certain behavior and regulates attitudes. It instructs and judges us from cradle to grave.

Traditions impose a certain style and structure on the family. They give a particular family a particular identity.

A tradition spells out the tolerances and intolerances of the family system. The fact that relatives are honored or shunned by other relatives usually has a great deal to do with any one member's emotional investment and expenditure in the family traditions. Heroes become heroes, and black sheep become black sheep, largely based upon how stringently they uphold family tradition.

Lynda relates, "After I did my genogram and thought about my family, what really stood out for me was a tradition of loyalty. What else could explain the fact that we carry on like crazy when we get together, yet we always get together?

"Every year, we go to Jamaica for Christmas. We were raised there, and my father still lives on the island. My mother and all of us (my four siblings and I) live in various cities on the East Coast, yet we fly to Jamaica and never consider staying here. My parents have been divorced for ten years, but my mother has never missed a Christmas there either. We all show up. And we all argue about what we want to do and complain about what we're doing, but there is no question in anyone's mind that the ritual of Christmas with Dad in Jamaica will be observed.

"My father is the traditional person in the family. He is Greek, and family means a great deal to him. I think my parents love each other and may get back together again someday. But my father felt 'betrayed' when my mother had a brief affair, and that's why he insisted on a divorce.

"My father's family never really approved of my mother. She was a New England WASP, and they were very critical of her always. Even now, my father hides from his sister the fact that he still visits my mother. Again, I think it's a loyalty issue. He probably feels that to still be involved with his ex-wife is somehow being disloyal to his family, particularly his sister, to whom he is closest.

"There are five children in our family, and I hadn't realized before I did my genogram that the family constellation is the same as my father's family: two males and three females.

"I'm really amazed to discover the label that really does describe our family. Loyalty certainly describes our relationships with each other—in many ways."

DIAGRAM 37 — TRADITION OF LOYALTY

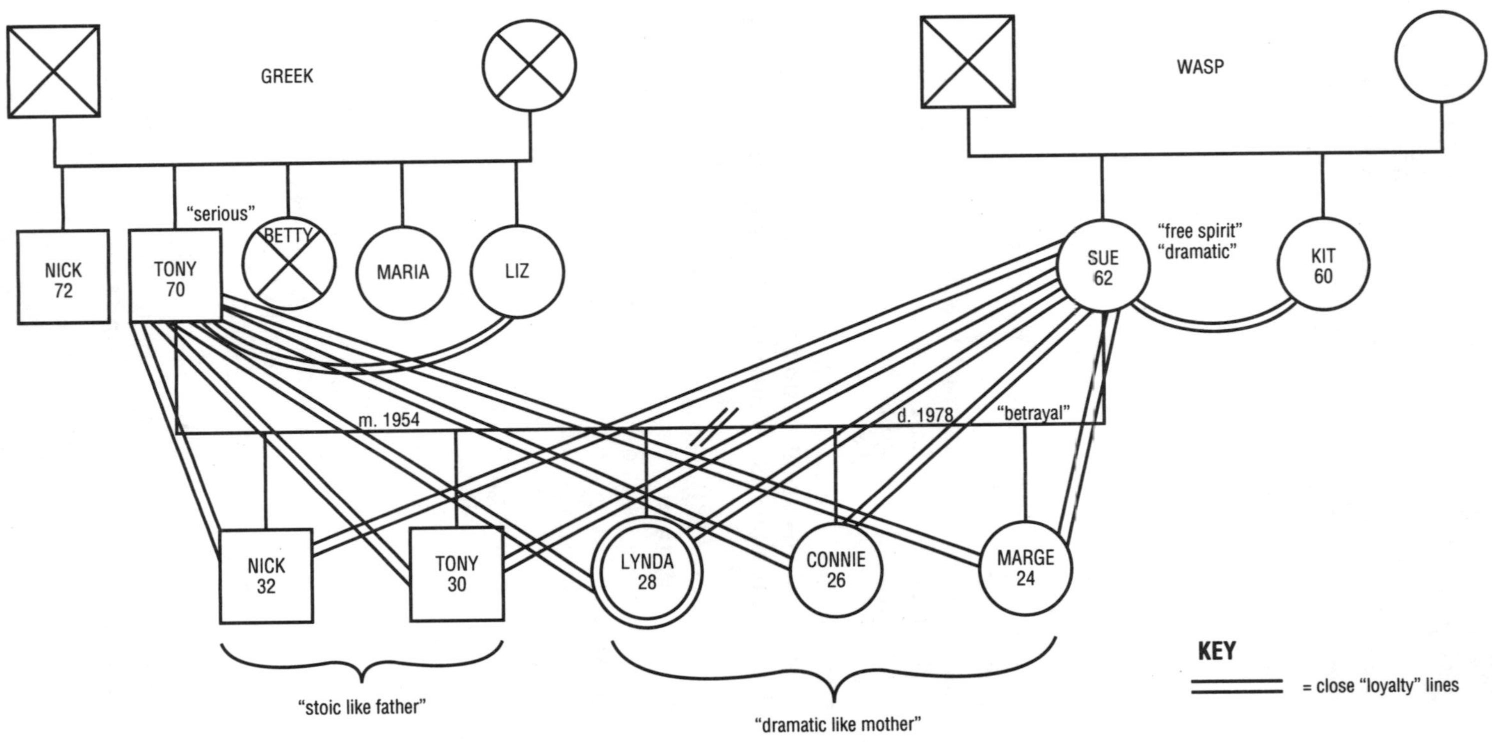

KEY

═══ = close "loyalty" lines

Lynda's genogram (Diagram 37) on which she saw loyalty as a family tradition is on the previous page.

Families That Protect and Overprotect

Some families have patterns of overfunctioning in certain areas. There may be a tradition of rescuing or overprotection. These characteristics show up in the major events in the family stories.

Sometimes people in dire straits need rescuing, and there is certainly nothing wrong with helping a friend or relative in need. But when the rescuing is out of proportion to the actual need or event, it usually means the rescuer is working overtime. Maybe he or she should look at his or her need to rescue. Is it being protective or overly protective? Is it fixing something for someone who would better benefit from fixing it alone?

If rescuing, or some other family trait, keeps showing up on your genogram, you can think about whether it is generally an asset or a liability. Don't be surprised if you don't get a clear-cut answer. Things are usually not so black and white. Consider the circumstances. What might seem like a virtue sometimes may appear as a vice at other times.

Leonard identified a family tradition of "save the children" in his genogram (Diagram 38). He comments, "My family stories are about rescue. There were several tales about child victims who were nonrelatives taken in by mother's family—the 'poor little match girl' scenarios. My grandparents took in other people's children long before there was official foster care. Their neighbors were killed in an accident, and they took in their two children and raised them as their own. There was no formal adoption; the orphaned children were just absorbed into the family.

"My mother and her older sister both had trouble conceiving, and it was only after they adopted children that they became pregnant. My aunt had two adopted children, then two of her own. My mother adopted my brother, Kent, and then had me.

"On my father's side of the family, his parents taught in a boarding school and were substitute parents for many kids. I never knew my grandfather, but all the family sto-

DIAGRAM 38 — SAVE THE CHILDREN

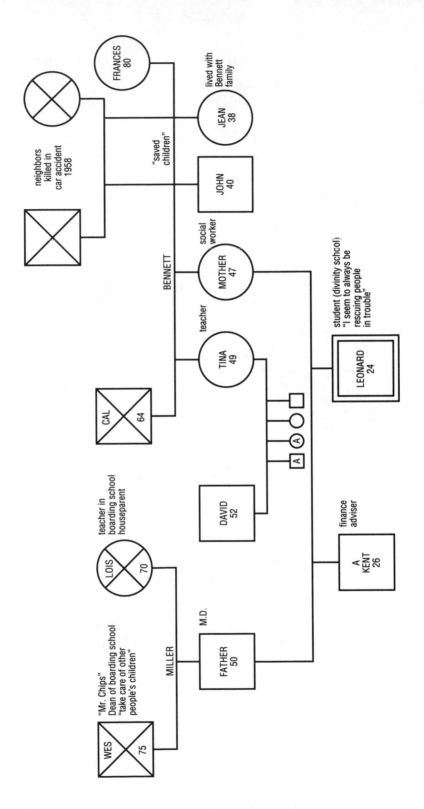

ries describe him as a 'Mr. Chips' type. My dad, not surprisingly, is an M.D.

"Thinking of my family stories made me look at my own behavior. I seem to be attracted to that wounded-bird type of a person who apparently can't survive without me. I guess the folklore of my family primed me for my rescue missions.

"Maybe now that I see how strong these family influences are, I can begin to focus on saving myself rather than some other poor soul. These patterns are tough to break, especially since everyone I know sees me as the protector and not as a person who sometimes needs protection too."

Families That Fight and Those That Don't

Many families have a tradition of either pacifism or fighting. Family stories may emphasize how a family member stood up for his rights or got the better of someone else. Other families pass down values of peacefulness and self-control.

Sylvia Foster outlined a pattern of "no competition" in her family (Diagram 39). The family rule seemed to dictate, "Don't fight." Sylvia felt this was a real disadvantage in her life and wanted to know how she could avoid this pattern and begin to fight and fight fair. She knew that it would be better for her to admit some of her negative feelings and that she was entitled to having her needs met in spite of opposition from her boss or husband.

"I hadn't thought about it before," says Sylvia, "but I wonder if the 'No Fight' rule in my family had anything to do with the fact that I married a police officer. My father was a police officer also. He retired from the force when he was forty-two and became an investigator for an insurance company. Two of my brothers are in law enforcement. There is a definite pattern in the family for following this line of work.

"My parents had eight children, and it seems ridiculous to me now that they would expect a bunch of kids *not* to fight. But competing or fighting simply wasn't allowed in our house. If we got into any kind of skirmish, we were sent off alone to our rooms. And, of course, we never

DIAGRAM 39 — DON'T FIGHT

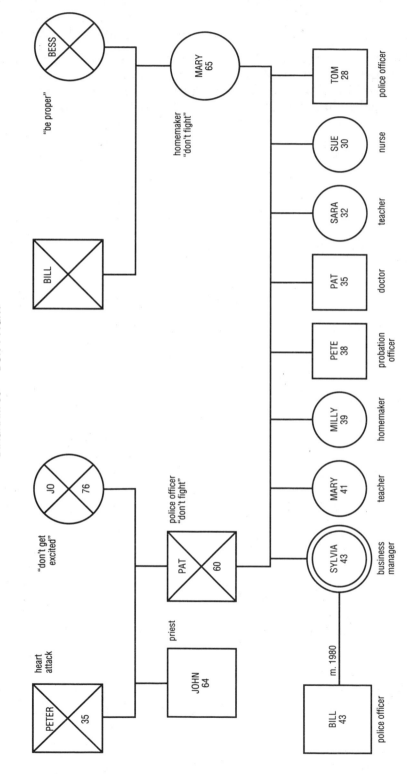

could get angry at our parents. We had to do everything our father's way. He was the law, and he maintained order at home.

"It is interesting to see the patterns of careers all of us chose. Most of us are in rather passive, caretaking, law-maintaining positions (education, medicine, and law). I'm the only one who chose business. Obviously, it presents problems because of my fear of anger and confrontations.

"My mother never raised her voice or shed a tear. She was always in control. With a stubborn and rigid husband and eight children, she must have felt anger and sadness, but you would never know it from her manner.

"The 'don't fight' rule was not all bad. In fact, when I discussed my genogram with my husband, Bill, we realized his family had the opposite rule. They bicker and fight all the time. I don't think that's great either. I guess that's the other extreme; neither one is good.

"In Bill's family, the constant fighting didn't really resolve anything; the fights just kept everything stirred up. It's still very hard for Bill to get a reading on what people really think. Just last week he received news that he made lieutenants' list. We visited his mother to tell her the good news, and she was thrilled. She even had tears in her eyes. But after we left her house, Bill asked me if I thought she was 'really happy' about his promotion. He still is not sure what she thinks or that she approves of him."

Family Jokes and Humor

Some family stories are humorous. If you see humorous themes in many of your family stories, you may want to make family jokes a subject of your genogram as Janice did in Diagram 40.

Janice relates, "Humor is a hallmark of our family, and it's helped us get through some difficult times. The two traditions about our family that I have always loved are that we shared a great fondness for animals and that we have a great sense of humor.

"We're still pet-crazy, and that's fine, but I am starting to question the humor. When I started to think about the things we laugh at or tease each other about, I realized

DIAGRAM 40 — HUMOR OR GALLOWS HUMOR?

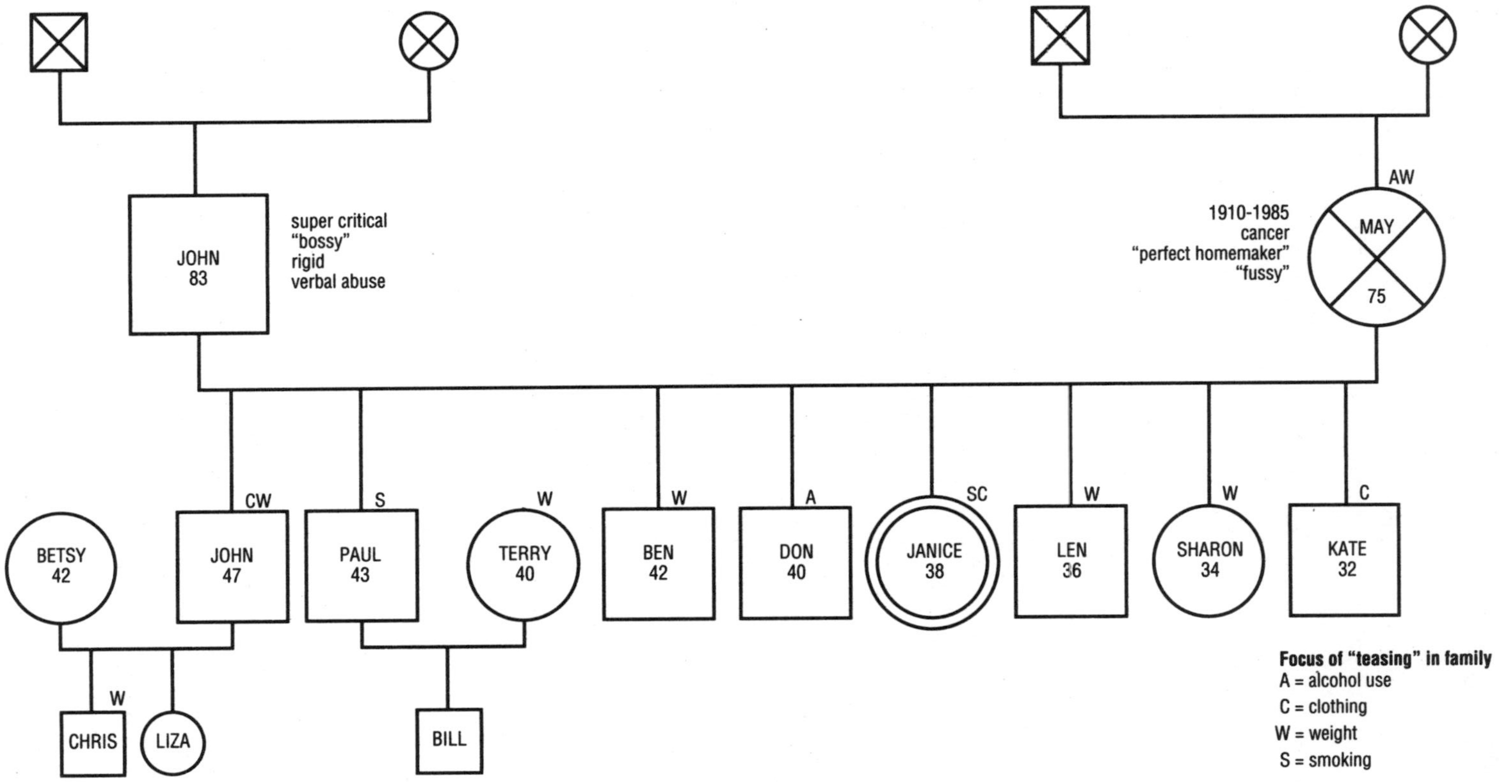

MAY
75
1910-1985
cancer
"perfect homemaker"
"fussy"

AW

JOHN
83
super critical
"bossy"
rigid
verbal abuse

KATE
32
C

SHARON
34
W

LEN
36
W

JANICE
38
SC

DON
40
A

BEN
42
W

TERRY
40
W

PAUL
43
S

BILL

JOHN
47
CW

BETSY
42

CHRIS
W

LIZA

Focus of "teasing" in family
A = alcohol use
C = clothing
W = weight
S = smoking

that very often we are picking at each other's vulnerabilities. Since there are eight children and three grandchildren, that's quite a bit of picking.

"Maybe our humor is not all that playful. Many of the jokes in the family are about people's weights, smoking, or how they dress. We make fun of each other's Achilles' heels. I know I've been very hurt by some of it, and recently my sister-in-law told me how depressed my brother John is now because of all the jokes the family makes about his weight. He's had a problem for years, and it's at the point where his health is really in jeopardy. But still the jokes persist.

"I was always ultrasensitive to my mother's criticisms, which were in the form of jokes. I was upbraided for being the kid who never could do anything right. She'd say things like, 'You're the sorriest excuse for a daughter I ever saw.' Or she'd say how 'dumb' I was playing bridge and other family games. My brothers teased me about being the 'runt' of the family, since I was the shortest and they're all six-footers. I caught a lot of grief from my two sisters about my smoking and the way I dress. They thought my clothes were weird and that I had no style at all.

"It's only now that I'm starting to realize how all this teasing may not have been meant to be malevolent but certainly did nothing to promote self-esteem. I don't think we mean to be harmful to each other, but the jokes didn't make me quit smoking and aren't helping John lose weight.

"I can see where we can use humor positively and negatively. I want to stop the gallows humor, which is basically not fun and is borderline abusive. As a family, we don't usually see where we are out of bounds and into someone else's space. We get so tangled. Maybe we can learn to separate a little and still be close without using 'funny' comments that only hurt.

"Maybe it's time for our family to talk about the 'jokes' that aren't so funny anymore, if they ever were. I see how my nieces and nephews are teased in much the same way. That kind of teasing isn't going to help their self-esteem either. Maybe we can break the cycle of unfair joking without losing our humor. I sure hope so."

Scripts: The Roles People Play

Most people get "tagged" early on in life. Family members tend to label us by their observations and expectations. Frequently we become the person they are anticipating. We behave according to our labels and generally believe them ourselves.

Common roles that seem to stick for a lifetime are family heroes, scapegoats, clowns, and invisible family members. Though these tags may never be mentioned in so many words, they can be powerful influences on the ways we think and act. Think about your family. Do any of these four roles apply to you or your siblings?

The hero does everything right and is glorified by the family. The scapegoat is the problem kid who gets a lot of family blame. The hero can do no wrong, and the scapegoat can do no right. Many families have a clown, the person who gets attention by being funny. Clowns also help to cover up family pain caused by family problems or people in conflict with each other. The invisible person is a family member who doesn't get noticed because he or she has given up trying to get any of the attention, which has gone to the hero and the scapegoat.

Favorite Sons and Daughters

The family hero is often the "favorite son" or "favorite daughter." Try identifying the "favorite" child (male or female) in your family and each of your parents' families. They may be difficult to spot, or they may become obvious when you draw your genogram, as they did for a young social worker, Frannie (Diagram 40).

Frannie has a younger brother who seems to be fulfilling all his parents' dreams. He is a doctor. She feels her own life is their nightmare, even though she is quite happy in her career. She thinks that her family doesn't even consider working with disadvantaged people a worthwhile profession, since she doesn't make very much money.

"Never, never does anyone in my family ask about my work," Frannie explains. "It really doesn't count to them. My brother's life is the only life my parents feel is worth examining. And, believe me, they examine it all the time. They have always been interested in his grades, his work, his girlfriends. I gave up trying to impress them a long time ago. They never ask me about my work or personal life."

Frannie's genogram (Diagram 41) showed that both of her parents were younger children with one older female sibling. Her parents were the family favorites also, and they had older siblings who were not terribly successful. Frannie's father's sister was a librarian (certainly not a limelight position), and her mother's sister had many emotional problems and committed suicide three years ago.

Frannie's genogram points to several other patterns. Her paternal grandfather's name was Francis, and she was named after him. Was she, she wondered, a disappointment to her parents from birth? Would her father (himself a "favorite son") have preferred a son to take on his father's name and carry on the "favorite son" tradition?

She also noted that both of her parents had conflicted relationships with their siblings, as she did with her brother. The age difference between Frannie and her brother was the same as between her parents and their older sisters.

Frannie has been able to find a few substitute families. She is in a women's therapy group, where she gets a tremendous amount of support and validation. She is engaged to Robert, a young man who comes from a warm and loving family. He has four sisters, all of whom are in careers highly respected by their parents. Frannie feels that she gets a great deal of affirmation from Robert and his family. Frannie has learned that even if she cannot get the attention she'd like from her family, she can get some of her basic needs met elsewhere. She says she now better understands the source of her parents' favoritism.

It should be noted that being the "favorite" child is not always a positive experience either. Guilt is often the heavy price the favorite pays for being singled out over

DIAGRAM 41 — FAMILY FAVORITES

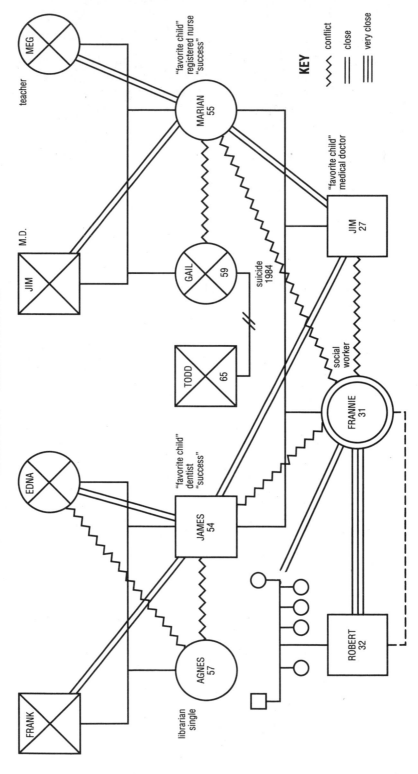

KEY

〜〜 conflict

‖ close

⦀ very close

brothers and sisters. It's not so easy to receive all the glory when someone else is receiving all the grief. Also, the expectations of the favorite are often too high. It can be a great burden for the designated special child to live up to the parents' extra-special expectations.

Family Troublemakers and Troubled Family Members

Are there people in your family who seem to have a black cloud overhead or who seem to be surrounded by troubles? If so, Jane's story may sound familiar to you (Diagram 42). She relates, "I have always been something of a 'Calamity Jane.' I was a sickly child and had every illness imaginable. I had pneumonia twice before I was a year old, croup repeatedly until I was six, an emergency appendectomy at eight, and a broken leg when I was ten.

"I was always getting into accidents and scrapes when I was a kid. If we went out to play on the swings, I was the child who got hit in the head with a swing. If we went for a camping trip, I'd come down with the measles. Even as I got older, I got into situations that caused my parents considerable grief. I was the teenager who got caught smoking in the girls' room or playing hooky. And then when I started to date, it was really wild. My parents never approved of the boys I'd bring home.

"When I was sent off to college, I think they must have breathed a sigh of relief. Actually there were no big traumas or incidents in those years. But six months after I graduated, I was married to a man twice my age. My parents were not thrilled. And Tom, my husband, had been married twice before and had a bunch of kids. He had three children from his first marriage, three stepchildren from his second, plus a daughter from that marriage. Overnight, I had seven children to deal with. Actually, it has been quite fun in spite of a crisis a minute. And I adore my stepchildren. Tom and I had a daughter together, bringing our brood up to eight by the time I was thirty.

"Looking at my genogram made me realize I'm not the only Calamity Jane in the family. My daughter seems to be a chip off the old block; she recently got expelled from high school. My father had his share of calamities when

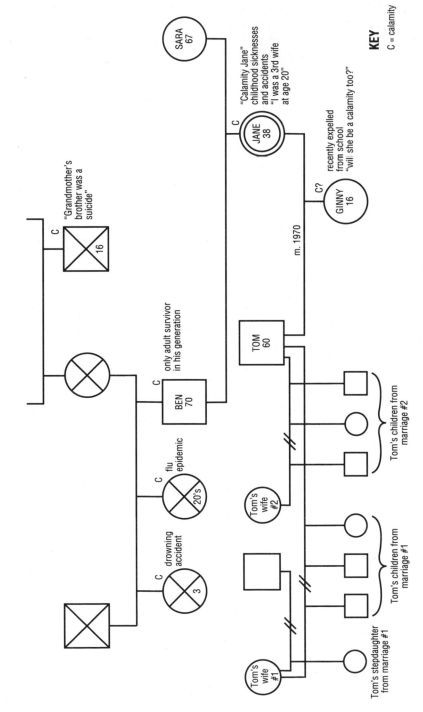

DIAGRAM 42 — CALAMITIES

"Grandmother's brother was a suicide"

C — 16

only adult survivor in his generation

C — BEN 70

flu epidemic

C — 20's

drowning accident

C — 3

SARA 67

"Calamity Jane"
childhood sicknesses
and accidents
"I was a 3rd wife
at age 20"

C — JANE 38

recently expelled from school
"will she be a calamity too?"

C? — GINNY 16

m. 1970

TOM 60

Tom's wife #2

Tom's children from marriage #2

Tom's children from marriage #1

Tom's wife #1

Tom's stepdaughter from marriage #1

KEY

C = calamity

he was growing up, too. His oldest sister drowned when she was three, and he had another sister who died when she was in her twenties. He also had an uncle who committed suicide. That was his mother's brother, but I never knew much about that tragedy. It was kept hush-hush. My grandparents are both dead, but I think I'll ask my father about my great-uncle the next time I visit him. I wonder if he was a calamity person, too."

Maybe there are members of your family who have other similarities in terms of what happens to them in life. Maybe instead of Calamity Janes, your family contains "Lucky Jims," people who come up smelling like roses no matter what happens. Try giving yourself a descriptive label and then see how many of the same labels turn up on your three-generation family tree. You probably will find roles and stories being repeated time and time again.

Families with Secrets

Perhaps the most tantalizing family stories of all are those involving secrets. Often people are never told important things about their families and only discover them through perseverance or by accident. If there are subjects or family members that are taboo in your family, or certain events that simply aren't discussed, chances are you have some family secrets waiting to be uncovered. Or perhaps, like Gene, you already know some family secrets that will be important to your genogram.

Gene says, "While I was drawing my genogram, I was reminded of several family secrets. The secrets involved my grandparents on both sides.

"It was only a few years ago that my younger sister told me one secret: that my father's mother (my grandmother) was never married to my grandfather. The other family secret involved my grandmother on my mother's side of the family. Her first husband was a womanizer and probably a bigamist. After he deserted my grandmother and their baby, he got 'married' to two other women and had several children with each of them. This all took place in a rural county in Oklahoma, and I doubt very much that grandfather ever bothered to get divorced. He was a 'rascal' and he didn't abide by many laws.

"My grandmother, my mother's mother, was someone the family never talked about much. There is mystery surrounding her death. I think she was shot to death in a barroom brawl, but I really don't know all the facts. I do know she was only forty when she was killed. My mother was really raised by her grandparents, who were very law-abiding and religious. My great-grandfather was a Baptist minister and very serious. My great-grandmother has always been more of a mother to my mom than her own mother was.

"Until I studied my family tree (Diagram 43), I never realized just how similar my parents' backgrounds were. It seems Mom and Dad were both abandoned by 'bad' fathers and rescued by 'good' ones. Neither of my parents were raised by their natural fathers. Both men left home when my parents were young kids. My grandmothers married men who adopted their children and took good care of everyone. My parents were both only children of their natural parents, and there were no more children out of those second marriages.

"There's another similarity between my parents. I think they are both alcoholics. I used to think it was only my mother who drank too much. She's always been a 'party girl.' But recently I've noticed that my father keeps right up with her even though he switches his drinks. He's just not as obvious. They've been married thirty-three years, and I think they are happy together, or at least they are good drinking partners. My father is a family man who never ran around with other women like his father. He always seemed ashamed of my grandfather's behavior.

"Both my brothers have family problems, and they probably drink too much also. They definitely have problems when it comes to relationships with women. My oldest brother, Milton, is thirty-five, and he's been married twice. He's known to run around, although no one really talks about this. My next oldest brother, Frank, who I have never liked, is a sicko, and I think he is abusive to his wife. He's always been violent, and my mother told me recently that my sister-in-law is not allowed to make any telephone calls (even to my mother) unless my brother is there.

DIAGRAM 43— FAMILY SECRETS

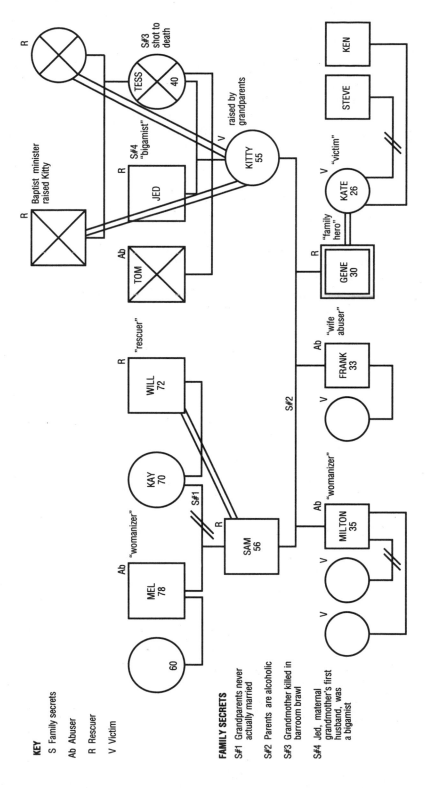

KEY

S Family secrets
Ab Abuser
R Rescuer
V Victim

FAMILY SECRETS

S#1 Grandparents never actually married

S#2 Parents are alcoholic

S#3 Grandmother killed in barroom brawl

S#4 Jed, maternal grandmother's first husband, was a bigamist

"My family problem today is that I don't feel as if I really have a family. I'm gay and not currently in a relationship. I work hard, but I don't have a social life. I'm pretty isolated. I think I spend too much time alone.

"The only person in my family I'm really close to is my kid sister, Kate, who lives in Hawaii. Obviously our telephone calls get very expensive, but we talk at least once a week. She's on her second marriage, and she's only twenty-six years old.

"There seems to be a pattern in my family of men not treating women very well, that is, either abusing them or abandoning them. A few of the men are rescuers, and I think I fall into that category when it comes to my sister. I'm also the guy at work who listens to the women who have problems.

"When I drew my genogram, I decided to put Ab over the men who were abusive and R over the men who were rescuers. With so many rescuers or abusers, it's not surprising that so many of the women in the family fall into the 'victim' category. I noted that role also on my genogram."

Gene discovered three roles in three generations and the family secrets on his genogram. In thinking about your own family, ask yourself whether you have relatives who tend to fall into the same roles. Like Gene, you might want to use family secrets as a starting point for your genogram.

Untangling Family Myths

Sometimes family stories and attitudes are more mythical than real, based more on what the family would like to be than on what it is. Family myths can be useful, at least temporarily, but they can also create tangled and traumatic relationships that are among the saddest and most difficult to understand.

Ida relates, "Drawing a genogram helped me to understand how complicated and tangled up our family tree is. For years we probably gave the impression of being one big, happy, blended family. My widowed mother had six children, and my widowed stepfather had seven children when they married twenty-five years ago.

"My father was an Air Force pilot who disappeared somewhere over the Pacific Ocean. Our stepfather's wife died during the birth of their youngest daughter. We may have looked somewhat like 'the Brady Bunch,' but we were not like that or any other all-American happy family.

"Even though thirteen little kids had lost a parent, those deaths or circumstances around those deaths were never ever mentioned. It was as if what happened never happened. We weren't allowed to talk about our dead parents or show any signs of sadness or longing.

"When I was in junior high school I had a photograph of my father in my bedroom, but it disappeared without a word of explanation. I found another photo, and the same thing happened. The unspoken message was that I should acknowledge only one father, my stepfather.

"As recently as a year ago, I had dinner with my stepfather and oldest brother, and I mentioned my father's relatives in a distant state. My stepfather mentioned that my dad was 'a great guy.' My brother was amazed. Until then, he had no idea our stepfather and father had known each other—let alone that they were pilots in the same squadron.

"That's just one example of the many secrets in our family. And the only reason I knew this was that I was the baby, my mother's only daughter, her favorite, and her confidante. She drank too much and took tranquilizers and would tell me a little about my father when she was smashed and melancholy. Of course, the next day she'd have totally forgotten she'd told me anything, so it was back to me pretending I didn't know anything. Because of my loyalty to my mother, I never repeated these stories to my brothers.

"Another myth in the family was that my father's relatives were snobs and didn't want to have anything to do with us after my mother married my stepfather. But last year, I flew out to my dad's hometown in Michigan and visited many of the relatives. Most of them were actually quite nice and told me what little they knew about my father. The only problem was that they had contradictory memories, so I didn't really get any clearer picture of who my father was. He's still a 'mystery man' to me.

"My stepfather was a war hero and saw himself also as a family hero, since he rescued my mother and together they raised thirteen children. But my mother died three years ago, and now he seems to be doing a real about-face. He acts as if he wants to forget his family. Right after my mother died, I visited him, and he had two huge portraits of both wives in his living room. Now he's taken them down, sent me most of my mother's belongings, and given away the dog they had for ten years. Does that sound like Father of the Year? But, in a way, it's typical of him. If he cannot change history, he pretends it never happened. Just as he always repressed his feelings about losing his first wife and my mother's drug problem, he is now acting as if he no longer has a family. He is gradually cutting off contact with all of us and becoming a loner."

Ida's genogram (Diagram 44) depicts the blended or "one happy family" that she now sees as an estranged family. Her current image does not live up to her old view. Does the reality of your family life jibe with some of your family myths? You might draw a "myths genogram" citing the family folklore and then draw another "reality genogram" to confirm or dispel these myths. How do they compare?

Families with Weak and Cutoff Roots

If you find that your family has few stories and very little connection with the past, this can tell you something about yourself as well. Consider Joyce's experience. She relates, "My genogram tells me a lot about my family because of what I don't know. For instance, I know nothing of my father's parents except their first names, that my grandfather was 'wealthy, rigid, and overbearing,' and that my grandmother was 'shy and rather psychic.' I don't know when they were born or died or anything at all about their families or their life.

"My father was born in 1920 and joined the Army when he was eighteen. After World War II, he never again contacted his family. I don't know why. He was the middle child and had a sister four years older and a brother a

DIAGRAM 44 — MYTH AND REALITY

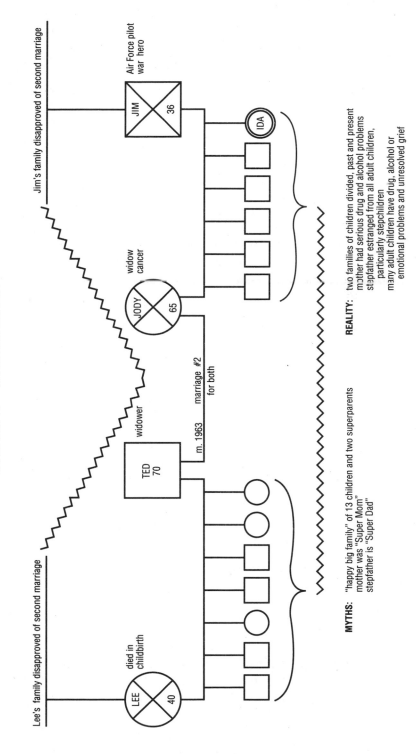

MYTHS: "happy big family" of 13 children and two superparents
mother was "Super Mom"
stepfather is "Super Dad"

REALITY: two families of children divided, past and present
mother had serious drug and alcohol problems
stepfather estranged from all adult children,
particularly stepchildren
many adult children have drug, alcohol or
emotional problems and unresolved grief

year younger. He said he never got along with his sister but that he was close to his brother who died of M.S. in his forties. But my father must have meant they were close as kids, because I don't remember my uncle's wife or his three children at all. Maybe I met them once or twice, but I don't even know their names.

"My mother wasn't much closer to her family, although I at least know the names of my mother's relatives. She was from a bigger family; there were four children, and she was the youngest. But she broke off all contact with her whole family after my grandparents died almost twenty years ago. Again, I don't know why. Also, her father was eccentric and left his family for a year to join a religious cult. This was when he had been married for some years and had teenage children.

"I guess I am following in their footsteps in terms of cutoff family relationships. My parents live only 100 miles from me, but I only see them at Christmas time. And my only sibling, my younger brother Ted, is cut off from everyone. He is a real loner; although he lives with my parents, he works at a very menial job and spends most of his time in his bedroom. Close family relationships were obviously not something my parents encouraged. Looking at their estranged backgrounds, I can see why.

"There are several other coincidences on my genogram. There was a pretty big age difference between my parents (twelve years) and between my maternal grandparents (fifteen years). And all of the women in my mother's family (her mother, two sisters, and herself) were alcoholics, drug addicts, or both. Mother stopped drinking twenty years ago, but both of her sisters died from drug- or alcohol-related illnesses.

"On a more positive note, I know there is a strong artistic streak in my mother's family. I'm an artist, and so is my mother. And her parents (who lived to be ninety-five and eighty-four) were artists also."

Joyce saw several patterns in the drawing of her genogram (Diagram 45). The artistic pattern made her feel proud of her gifted heritage, but the estrangement pattern made her question how much she was contributing to the distance she had with her parents and brother. She can

DIAGRAM 45 — ART AND ALCOHOL

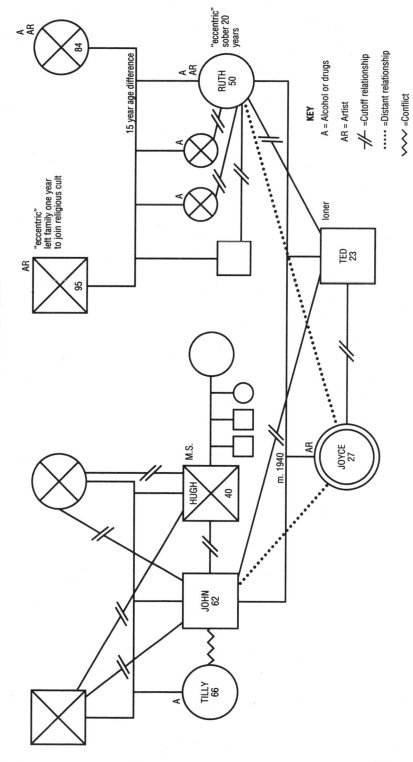

KEY

A = Alcohol or drugs

AR = Artist

‑//‑ =Cutoff relationship

...... =Distant relationship

〰〰 =Conflict

decide that it is in her best interest to maintain this distance, at least for now, or she can decide to participate more in the family.

6
IMPORTANT COINCIDENCES

The Amazing Power of Events

One particularly fascinating aspect of genograms is their capacity to link simple facts and data to form revealing patterns. Often family events can seem at first to be simply coincidences. For instance, in one family, the oldest sons in three generations are all doctors. In another family, a child named for a relative who's considered eccentric and something of a family rebel turns out to have the same "rebel" status in her generation. Often, certain seasons or holidays seem to be sources of discord, depression, or momentous events for certain family members. A genogram can reveal many reasons for these seeming coincidences. It can show patterns of success, health, marriage, and emotional states and, most importantly, point to the *meanings* behind them.

Some patterns to watch for as you draw your genogram are birth order, careers, names, marital status, and important dates and anniversaries. By charting these facts on a genogram, you will better understand your family's past and gain valuable insight into your present and future. This chapter shows how to interpret some of these facts. You will also see what kind of important coincidences to watch for as you do your family genogram.

Birth Order

Birth order is position in a family by age. We know that certain patterns of behavior are related to birth order. Studies have shown, for instance, that firstborns tend to be leaders, perfectionistic, independent, authoritative, successful, and aggressive. Middle children tend to be mediators, tenacious, underachieving, and unpredictable. Babies of the family are often charming, manipulative, comic, demanding, and dependent.

You've probably heard the expression, "He's a typical middle child." The person is referring to the person born in the middle of the family. Typically, the middle child doesn't attract as much parental attention as a first child or the baby of the family. He often ends up feeling "lost in the shuffle." When someone says, "She's the 'baby' of the family," the reference is not simply to the fact that she is the youngest. There is a "spoiled child" connotation to the word baby—an expectation of preferential treatment.

We assume people will have certain characteristics based upon chronological place in family constellations. Labels are often wrong and unfair conjectures, but labels attached to birth order provide general classifications which may help to explain certain behaviors. But be prepared for exceptions to the rule. For instance, a male middle child between two sisters may be "special" because of gender rather than birth order. His parents may expect him to be a leader (like an eldest) and at the same time be very indulgent and treat him like a baby. This middle child then doesn't get lost at all. Instead he gets all the positive attention usually received by both the eldest and youngest children.

When there is more than five years between two siblings, the younger becomes more like an only child (typically, the only child receives a great deal of parental attention and is, consequently, accomplished, poised, and articulate, though perhaps a bit spoiled). If other children follow, he or she then becomes an "oldest" to the younger siblings. Sometimes there is a big age gap in families that creates separate families within the family. For instance, the genogram of the Prince family (Diagram 45) shows two sets of three children each in one family system.

Jack, Jim, and Bob Prince were each two years apart, and then there was a gap of seven years before Lilly was born. Her birth was followed, two and four years later, by those of Tom and Jan.

Jan says, "By the time I was in first grade, my three oldest brothers were away at boarding school or college, so I really did not see much of them after that. It often felt as if we were two separate families with the same parents. Even today I am quite close to Lilly and Tom but only see my big brothers at family weddings and funerals."

Even though Lilly is the fourth child, she feels more like the eldest of three siblings. She is more of a "big sister" than a "little sister."

The general characteristics based upon birth order only partially explain personality makeup. Many other family tendencies can certainly affect behavior. Other relevant factors are sex, number of years between siblings, mixture of males and females in the family, physical and emotional traits, relationships between the parents and between the parents and their siblings, and the birth order of the parents.

Dr. Kevin Leman, in his book *The Birth Order Book: Why You Are the Way You Are,** cautions, "Birth order is never a final determinant of anything, but it is an indication of problems or tensions that you might discover—or create for yourself—as you go through life. . . . Knowing the characteristics of your birth order and knowing yourself better is one of the first steps to learning how to get along with your mate and building a happy life together."

Birth order gives clues rather than pronounces indictments. Following are some clues birth order suggests about human behavior.

---

* Old Tappan, N.J.: Fleming H. Revell Co., 1985.

DIAGRAM 46 — TWO FAMILES IN ONE

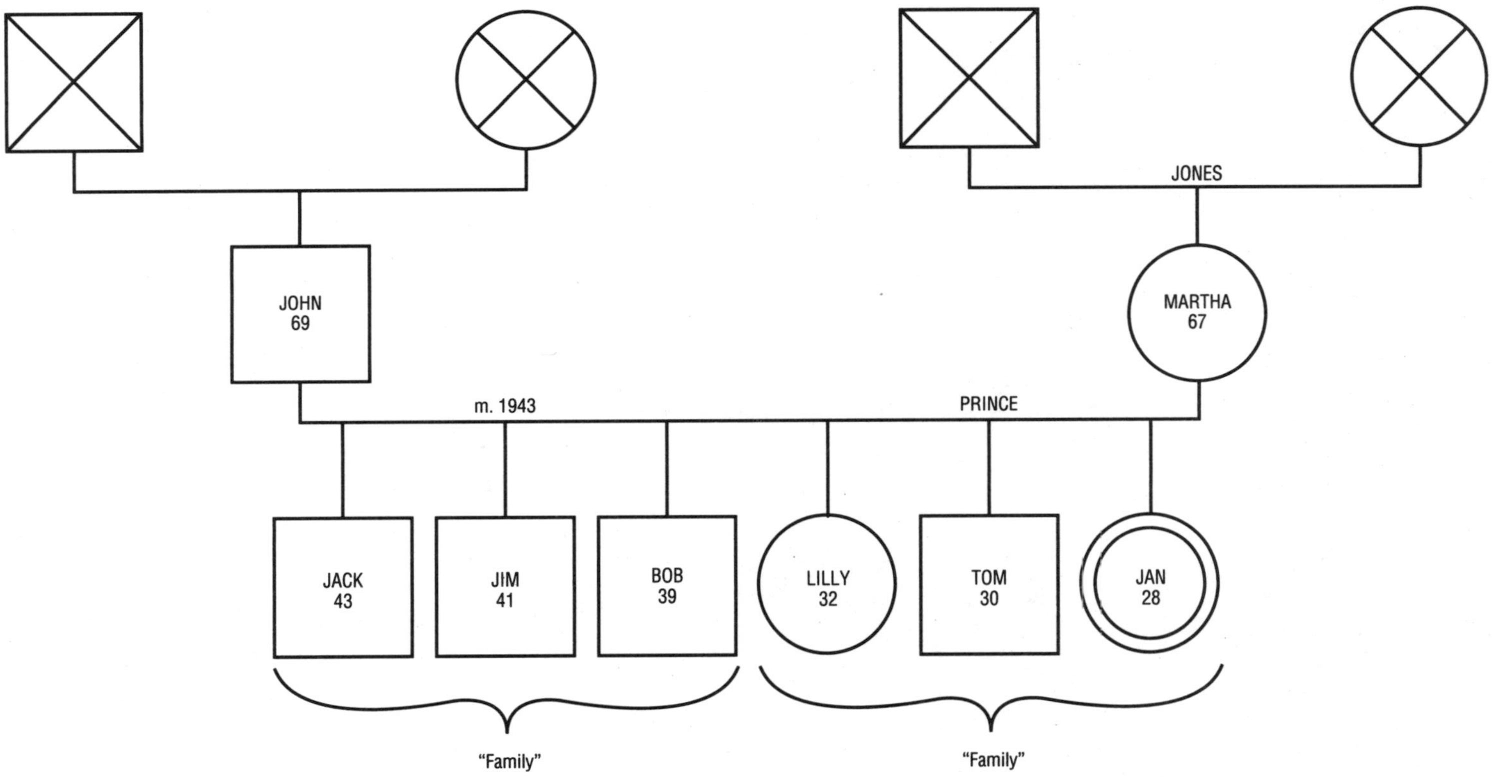

- Birth order explains, for example, why your oldest sister, Jane, is a perfectionist, why middle brother, Joe, is a show-off, or why Aunt Nellie's nickname was "Babe."

- Birth order sometimes explains why people react to situations in certain ways—for example, why Mary (an eldest) is upset, Johnny (the middle child) is complacent, and Suzie (the baby) is oblivious upon finding out that Dad just lost his job.

- Birth order may explain career choices. For example, most doctors and lawyers are firstborns, while entertainers are often babies of the family.

Consider how your birth order may explain how you fit in your family of origin or how you interact with your relatives and friends. If you're an eldest, for instance, do you like to be "the boss" at home, at work and in your social group?

By looking at birth orders on a genogram you can make certain assumptions and then check them out. For instance, if both parents are eldest in their families of origin, they may overidentify with their own eldest child. Consequently, that child may get considerable pressure to be "the best." The parents may project their need to succeed onto the child they assume is most like them.

However, consider an example in which parents are most emotionally connected to a disabled second child. In this case, regardless of birth order, the second child will be the favorite and, perhaps, the most indulged. Chances are that he or she will be the most immature or childish.

The birth order of couples often, positively or negatively, affects marriages. If, for instance, both partners are eldest children, they may both want to be "the leader." Obviously, this could create problems. If both children are middle children, they both may fear and avoid confrontations, which makes it very difficult to have open and honest communication. If both partners are babies of their respective families, they both may expect to be indulged by their mate and become resentful when they're not spoiled.

Keep in mind, however, that good marriages work only when couples, regardless of family birth order, work at the marriage. Caring, consideration, support, respect, and good communication are some of the areas requiring ongoing work. These principles also hold true for other family relationships.

Same Time Next Year: Anniversary Reactions

In many families there are coincidences in which significant events are strikingly similar in time, place, or circumstances. When family members react to the fact that a particular date is the anniversary of some special or traumatic experience, this is called an "anniversary reaction." The three-generational genogram often makes clear a connection between present depression or grief and the anniversary of a previous loss or losses. For instance, a person who feels blue in August may note on his genogram that his beloved grandmother died in that month several years before.

Often family members don't make a conscious connection between depressions and the anniversary date of some critical or traumatic family event. The genogram highlights these events and helps us to understand certain reactions. It is quite common, for instance, for a child to be unhappy approaching the same age as that of a parent's death. Suicides and "accidents" are often intergenerational, with deaths occurring at the same time and by the same method. Sometimes people commit suicide on the birth or marriage date of a close relative.

We can often predict or anticipate certain types of behavior when a particular life cycle event occurred in earlier generations and may then be anticipated in the present one. If, in several generations, there was a pattern of mother's death or critical illness in childbirth, expectant parents in the following generation would probably experience a great deal of anxiety. Family members would be afraid that history would repeat itself.

Seeing anniversary reactions on a genogram can alert you to possible high anxiety or pressure regarding coincidences in date, age, and point in the life cycle. Shirley Bates discovered such a coincidence on her family genogram (Diagram 47). Her paternal grandmother was two years old when her mother died in childbirth. Shirley's mother was two years old when her mother died of tuberculosis. And Shirley has a daughter, Elizabeth, who is a year and a half old. Given the two family incidents of death at the same time in the life cycle, it is understandable that Shirley is preoccupied with her health at this time.

Bob Green also reports an anniversary phenomenon in his family: "My great-grandfather and grandfather both committed suicide on their fiftieth birthdays. And when my father was approaching his fiftieth, he had a severe depression that required a six-week hospitalization. He is fine now, but he was very depressed for about a year."

Another anniversary reaction occurred in Betsy Little's family: "My grandfather died the day after he retired. He was sitting on a park bench, and he had a heart attack. His father had died in much the same way. He was in his sixties and was sitting in his yard when he just keeled over."

A less dramatic anniversary reaction was noted in Phyllis's story: "My mother had her first child, me, when she was thirty-five years old. I had my first child, a daughter, when I was thirty-five."

Significant Separations

When a loved person is lost, by death, separation, divorce, or desertion, people left behind will probably be very lonely and may choose to be in a kind of social isolation for a certain mourning period. However, it's not a good idea for anyone to stay in isolation. People need people. And family losses often means people will have to look beyond the family for the love and support they need.

It's important to think about whether being a loner or being lonely is the result of circumstance or of self-imposed isolation. Usually, it's the latter.

Reaching out to others is not easy for those who take pride in being self-sufficient, have grown accustomed to

DIAGRAM 47 — MATERNAL DEATHS

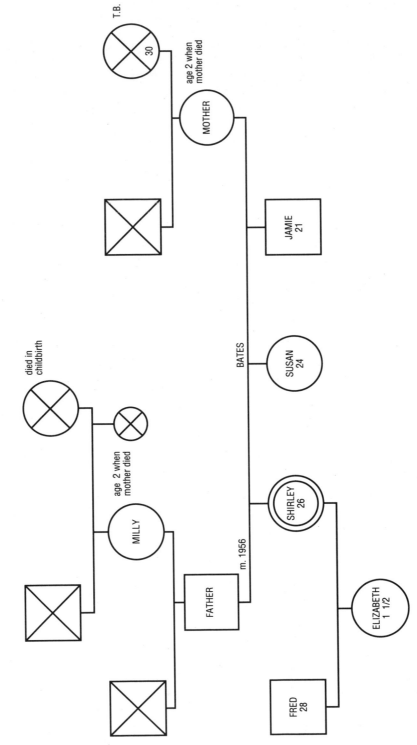

being fiercely independent and view a loner status as a positive family trait. But when the loneliness and sadness persist for a long time, it might be wise to see how other loners in your family managed their isolation. A family pattern of separation or loneliness may indicate a need for family members to learn how to cultivate relationships more successfully.

Helen is twenty-six years old and single. Her genogram revealed a pattern of women as loners (Diagram 48). The men in the family quickly disappeared from the family by dying or deserting at young ages. Helen's father deserted the family when Helen was four years old: "He left, moved across town and started a new family. I never saw him after that. He died three years ago." After her husband's desertion, Helen's mother raised four children on her own. She never remarried or became romantically involved with another man. She has "no social life at all."

Helen's maternal grandfather died when he was in his forties and Helen's mother was twelve. Her grandmother never remarried.

Helen's maternal aunt and uncle each live alone. Her uncle never married. About her aunt, Helen says, "Aunt Bea's husband died when I was thirteen. He had a heart attack at fifty-one, leaving my aunt and two girl cousins. All three women are single."

Helen's next oldest sister lives alone following a divorce after three years of marriage. She has no children. Helen's oldest sister is married with three children. Her husband is severely depressed (and thus emotionally unavailable to the family). Again, men in the family seem to be "missing" in one way or another.

Eight years between Helen and her closest (in age) sister, Liz, and seven years between Liz and the oldest sister, Elaine, have contributed to lonely childhood experiences. Although Helen was the youngest of four and the baby of the family, she felt like an only child because of the big age differences that separated her from her older siblings.

These types of significant separations, repeating through generations, are another not-so-coincidental coincidence to look for on your genogram.

DIAGRAM 48 — DESERTION AND ISOLATION

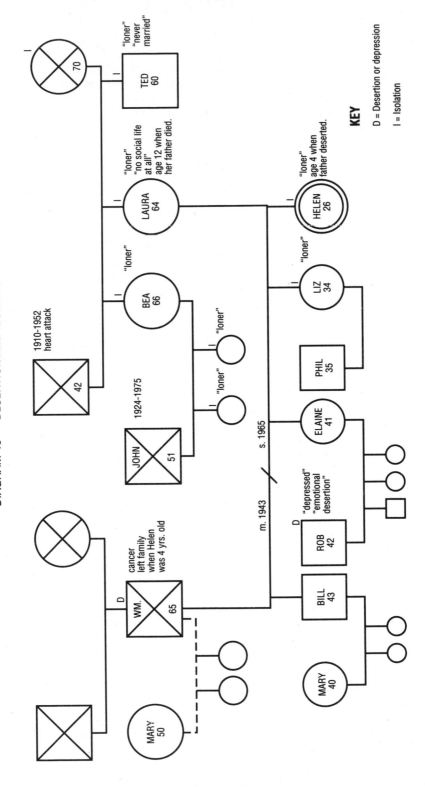

KEY

D = Desertion or depression

I = Isolation

The Nonspecific Genogram

One of the most fun genograms to draw is the "magical mystery tour," or nonspecific genogram. Don't look for anything specific in this uncharted map. Just draw your genogram and put down whatever comes to mind. Sometimes the facts speak loudly for themselves.

This kind of genogram allows you to be an explorer of new territory. You have some idea of where you might like to go, but you're not quite sure how you will get there, and you certainly have no idea of what you will find at the end of the journey. It's boring to always take the safest and fastest route. Now you have the opportunity to follow the uncharted territory. You get to take the little dirt roads instead of the superhighway.

Take your time, and see where your pen and mind lead you. If a particular date, situation, or event intrigues you along the way, check it out further. See what becomes a pattern rather than looking for a definite pattern you already know about. There are always surprises in family genograms if you allow yourself to be surprised.

Lee decided to do a nonspecific genogram (Diagram 49). As she was charting people and their occupations, she began to see a tradition in her family of "owning your own business." Her father had a small business, which he managed with the help of his wife. He had recently convinced his son, Ben, to take over the family business. To do this, Ben was going to give up a very good job in a Fortune 500 company.

Lee's paternal grandparents owned their own business, too, a pub in Ireland. Her mother's brother prided himself on "never working for anyone" and had stayed on the family farm in Scotland, where he continues to raise sheep.

Looking at these facts on her genogram, Lee began to understand why she was increasingly unhappy in her career as a lawyer in a large corporate firm. She had just received an offer to go into a partnership with a friend in a new field of law, and she felt conflicted. She also felt some anxiety about how her family would react to her leaving a secure position to go into something risky. Her genogram reassured her that she could approach her

DIAGRAM 49 — FAMILY BUSINESS

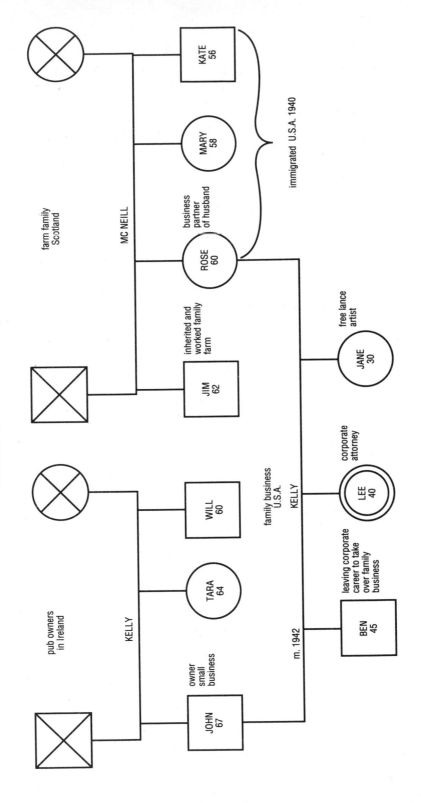

family by pointing out her desire to follow in the family tradition of taking charge of one's life in the work arena. Lee received some comfort from the realization that many people in her family had taken risks and been successful. She was still unsure what she was going to do, but the genogram exercise seemed to pull her more in the direction of taking this major leap from the known to the unknown.

As this example shows, while some of the important coincidences you find may be ones you look for, other patterns will take you by surprise.

Is history repeating itself in your life? Have some of the coincidences in the areas of birth order, anniversary reactions, particular patterns—even actual events or personality traits—been part of your genogram? Flushing out the coincidences on your genogram can alert you to study your family beneath the surface of those coincidences. What kinds of belief systems, traditions, actions and attitudes led to these coincidences? Do you see yourself following in other people's footsteps? If you do, you can choose to keep following the prescribed path or decide to lead yourself off the family trail into another direction.

7
GENOGRAMS FOR PROBLEM SOLVING

Mission Possible and Productive

When you begin your genogram, you may be hoping for the answer to a specific question or problem. This is an excellent reason for doing a genogram, but you should keep in mind that the genogram exercise usually raises new questions as it answers old ones. Often people see familial patterns that give them pause; part of the answer can be simply pausing to consider a new way of looking at things. If you feel you are at a crossroads, a genogram often flashes a caution light, signaling that this is a place where you should stop and consider where you really want to go and the best way to get there.

When Janice Davenport began drawing her family tree she was engaged to be married, and she was having serious misgivings because of her fiancés drinking behavior. He was having problems on his job, which seemed to be connected to his frequent nights out with the boys. He was just not getting to work on time or doing his job well.

Janice knew Sam was a heavy drinker, but she was beginning to question the seriousness of the problem. Neither of her parents drank, and she drank a cocktail only on special occasions. What, she wondered, was she doing marrying someone with a possible drinking problem?

In an effort to find some answers, Janice decided to telephone her parents who lived in another state so she could begin to work on her genogram. She was curious about the kinds of marriages people in her family had, and about relatives who "drank too much" or were involved with mates with drinking or drug problems.

Initially, Janice was reluctant to place the telephone call to her parents. She thought they would find it difficult talking about themselves and their families. But she called anyway. She explained that she was interested in learning more about her history. At her mother's suggestion, her dad got on the extension, and they had a lively three-way telephone conversation.

By the end of the hour-long call, her parents told Janice that they found the talk interesting. Janice felt closer to them and was determined to learn more. She was surprised and pleased to find out that they were much more receptive than she had thought they would be.

Old News That's New

Janice did, in fact, learn a great deal about her family (Diagram 50). There were some small surprises and several big revelations. In her words, the two things that surprised her most were: "that my paternal grandmother 'hated alcohol' yet married an alcoholic," and that "my artistic interests are explained by a creative streak in my family that I was never aware of."

Janice's mother, Jane, described herself as someone who "loves to travel, meet new people, with a passion for music." She talked of being an accomplished pianist at an early age, graduating from high school at age sixteen, and having a short, but successful, career as a dressmaker and draper. Janice's father pointed out that his wife is also "a wonderful cook, especially in the baking department" and

DIAGRAM 50 — ALCOHOLISM

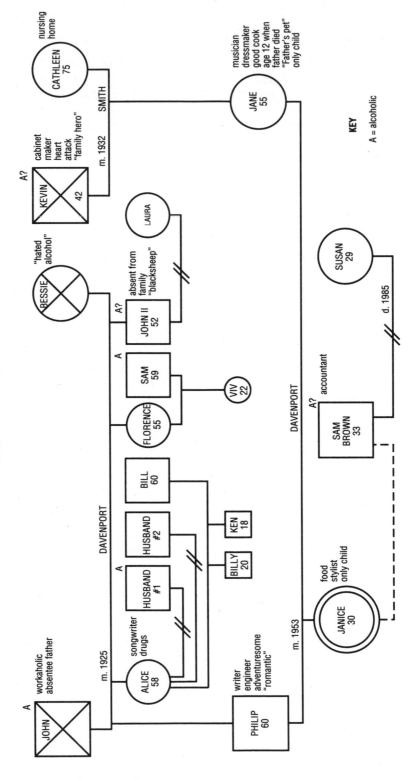

KEY

A = alcoholic

that she "manages money very well."

Janice's father, Philip, saw himself as "someone who loves cars, has a sense of adventure, and is a frustrated writer." He spoke of how much he had loved working in the public library as a teenager. He had been an avid reader, and the job was more a home away from home for him. His wife added that he is also a very good writer, and she reminded him of the scores of poems he had written during their courtship.

Janice had always viewed her father as a no-nonsense, practical, hardworking engineer. She was delighted to find out he had been such an intense and romantic young man. The conversation also reaffirmed her belief that her parents basically have a good marriage. They share a great deal in common; they are creative, practical, and have a sense of adventure. For instance, they talked excitedly about the cross-country trip they made in 1950 with their six-month-old baby (Janice) in an old car that broke down every other day. That love of exploration has continued throughout their marriage. Afterwards, Janice realized that even the telephone call was characteristic of how accepting they are of new challenges and how the spirit of their cooperation allows them to take risks and to have good times together.

On the not so positive side, it also became clear that her parents have suffered significant losses and that these memories are still very painful for both of them. Janice's maternal grandfather, Kevin Smith, a skilled cabinetmaker, had died of a heart attack in his early forties, leaving behind a wife and twelve-year-old daughter. Jane was devastated by her father's sudden death, as she had been very close to him. "I used to go everywhere with my father," she told Janice.

According to her father's report, Janice's paternal grandfather, John Davenport, was an alcoholic and workaholic. "He was just missing most of the time," Philip said. Both her parents feel sad about the parenting they had longed for and never had or lost.

Chemicals and Creativity

Philip Davenport also talked about his two younger sisters and his brother. His sister Alice, who is two years younger than he, has a serious drug problem. She has been on prescription medications for anxiety and insomnia for twenty years. She is now in her third marriage, and she has moved back and forth across the country several times.

Janice's Aunt Alice, in spite of her problems, is a talented songwriter, and several of her songs were published ten or more years ago. Janice recalled that when she was growing up Aunt Alice was her favorite aunt. Janice was drawn to her aunt's creative and free spirit. Considering this later, Janice had to question her aunt's chemical dependency. Was that, like Sam's drinking, also a draw?

Janice's father's second sister, Florence, married a man whom her father described as a "boozer and troublemaker." Philip reported, "Florence denies everything; she is in a world of her own. She thinks all of her husband's problems are problems due to bad bosses, bad luck, and bad times." According to Philip, his sister refuses to see that both her husband and father have or had serious problems related to their drinking. Even today, Florence speaks only glowingly about her father, and she is furious when her brother or sister refers to his drinking and irresponsibility.

Secrets

The last relative her father talked about was his brother. There is much mystery about her father's youngest sibling, John II. Only after persistent questioning did Philip acknowledge that John II is missing from the family. He had been in all kinds of trouble as a teenager, was married and divorced by the time he was twenty, and had been in and out of institutions (jails and hospitals). His present whereabouts are unknown. No one in the family has seen or heard from him in over twenty years. Dead or alive, he is the black sheep of the Davenport family.

Like Janice, her mother, Jane, is an only child. Both women had felt very lonely growing up and had always wanted a baby brother. Knowing they both had shared the same wish made Janice feel particularly close to her mother.

Janice learned a great deal about her family in this initial telephone conversation. It answered some questions and prompted even more. She continued her research on her next visit home, and this is when she learned from her mother the truth behind an old family myth.

Myths

Besides family secrets and black sheep, most families also have myths. The big myth in Janice's family was that her Grandfather Smith was a hero, not, as the facts seem to show, an alcoholic. It was only days after her phone call with Janice that Jane began to think about the glorification of her father and to question the cause of his death, an early and fatal heart attack.

On her next visit to her seventy-five-year-old mother Cathleen, in a nursing home, Jane tried to learn more information about her father's reputation as a man who loved a good time. Cathleen admitted that her husband Kevin had, in fact, "loved a good party."

Just what were these parties? Jane learned that Kevin's father (Janice's great-grandfather) owned a bar, and every day after work, Kevin would stop in to have "a few drinks." Often these few drinks turned into a "party" and Kevin did not return home until the party was over, that is, when the bar closed at 2 A.M. Jane began to remember the many, many nights she and her mother would anxiously await his return. She also recalled that loud arguments often followed those middle-of-the-night scenes.

Yet there was an unspoken agreement between Jane and her mother, from then until now, never to mention Kevin's drinking behavior. They always talked about his outgoing personality and the fact that he was a "good provider" and an "excellent craftsman." Until now, wife and daughter preferred to remember Kevin with no flaws. Until Janice's phone call, her mother had blocked all painful memories of her father as a drinker.

Making Changes

Some of this information helped Janice in her present predicament. We know that most people do not make startling discoveries about their families and make immediate changes in their personal lives. However, Janice had definitely made some observations that she felt she could not totally ignore. She says they are "food for thought" and make her feel less confused and angry. She certainly is not ready to break off her engagement to Sam, but she is prepared to consider taking steps to educate herself about alcoholism and the possibility that she plays a part in his addiction.

Janice decided to try to talk to Sam about her fears. She decided to suggest premarital counseling to see if he was willing to learn more about the ways they were relating and to try to improve their communication.

Reflections

Janice believes she does not have to immediately act on any of her discoveries, but she intends to reflect upon them. She made the following list of family tree facts that seemed particularly meaningful to her:

1. Grandmother Davenport "hated alcohol" but married an alcoholic.
2. Grandfather Smith may have had a drinking problem.
3. There are conflicted opinions and considerable denial about drinking problems of people in my family.
4. My parents have a good marriage; they are committed to and supportive of each other. They share common values and interests.
5. I'm pleased to learn about the creativity of many people in my family.
6. My family has a "hero": Grandfather Smith.
7. My family has a "black sheep": Uncle John.

8. My family has secrets, i.e., that Grandfather Smith had a drinking problem, that Uncle John is missing.

9. My family has a history of hard work and creative talents.

10. My family is probably not so different from other families.

You might consider making a similar list of your discoveries and conclusions when you do your family tree. Your observations most likely will not point to an immediate and simple answer to a particular problem, but they will provide valuable insights you can use and build on.

Although Janice's initial motive for drawing her family tree was to investigate issues of substance abuse and marital relationships in the family, the outcome was informative on even more than these two subjects. By speaking directly and honestly, Janice opened up communication with her parents, and this had many side benefits. In addition to finding out a great deal about her family and herself in the process, she had a very strong feeling that this kind of communication was only a beginning. The initial telephone conversation with her parents was much more meaningful than she had expected. Like Janice, you may find that discussing family history openly and honestly is a very important ingredient in promoting healthy family relationships.

Taking A Stand

Janice's friend Larry's genogram prompted him to take more immediate action than Janice did. A twenty-eight-year-old physical therapist, Larry recently began to question the competitive spirit in his family. His youngest brother had just won an international scholarship award, and Larry was displeased with his initial feelings of envy and jealousy. He also thought about the family hoopla over the award. He thought about how his family valued "achievements" rather than people and their personal worth and "feelings." Everything was measured in terms of "success" in winning yet another "trophy."

"There are eight of us in the family, and every single one of us has always been a high achiever. We felt we had to excel in everything: school, scholarship, sports, popularity, music. That was the way we got our parents' attention.

When I was in high school I started to see the absurdity of all this when I realized I was lobbying to get on the Principal's Committee just to chalk up yet another honor. Already I was class president, at the top of my class scholastically, in the debating society, on the basketball and swimming teams, and band captain. Why did I have to add something else I wasn't even interested in? I was always looking to do something my brothers and sisters hadn't done. I began to realize how there was a kind of "moving trophy" that all of us vied for and went from person to person. All the attention was on what we did, not who we were.

"My family is not the 'perfect family' and I want to have a more balanced view of myself and of them. I don't have to be 'a credit to the whole family,' I can just be me and try to accept other people for who they are. I started to practice this over the weekend when I spoke to my brother Pat who just won the Rhodes Scholarship. I told him that I thought he should do what he wanted to do with his award, not what everyone in the family thought would be the next Big Achievement. I told him he was important, not the award. I felt good about the conversation because I really tried to put myself in his shoes. Later he came back to me and told me how much he appreciated what I had to say.

"Success is fine. Obviously the family values of reason, leadership, and achievement paid off in many of our lives. But what I see missing on the genogram (Diagram 51), and in our lives, is an appreciation for who people really are. There never was an opportunity to talk about feelings.

"I was so concerned with pleasing my parents and competing with my brothers and sisters that I failed to notice who I was or who anyone else was. I guess I want to stop being the rule-book person and play by my own rules. I want to find out what I'm feeling and what I've been protecting myself from with all those activities that just upped the odds. The ladder was always being extended

DIAGRAM 51 — ADDICTED TO SUCCESS

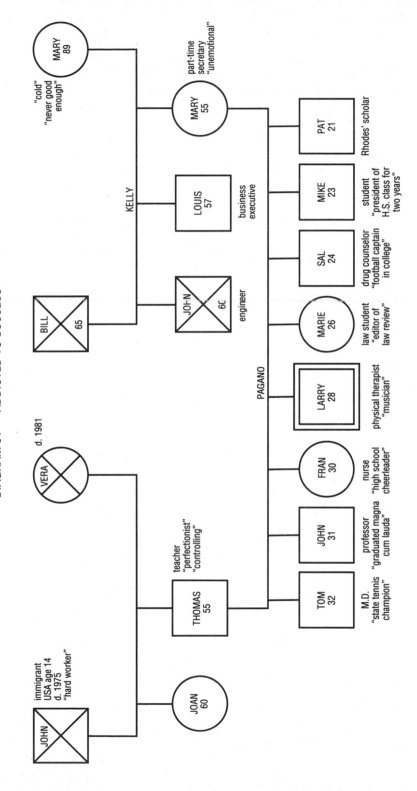

with yet another rung. I can stop thinking about what next great thing I have to do to get my parents' attention. I can begin doing what pleases me."

Larry's story is a good example of how doing a genogram can provide the insight needed to make changes. Most people, of course, don't act immediately on insights. But the kinds of revelations a genogram provides often initiate a thought process that leads to behavior and attitude changes.

If light bulbs go off for you when you do your genogram, be grateful for these revelations. But be cautious as you learn new information. You may need time to digest this material before you make decisions or take action.

Even though you may be tempted to share your new insights with family members, you may decide not to. Just because you may feel ready to deal with some tough revelations that may change your family view, doesn't mean other members are. Your interpretation may be skewed, too. It's one thing to mention the genogram to others and see if you elicit interest. If your sister asks, tell her your theories. It's another to diagnose everybody!

If you don't like everything you learn about your family, you may want to lash out or blame people for things that they may have said or done, intentionally or unintentionally. But blaming or attacking people who may not share your conclusions will probably lead to troubled family relationships. You're trying to mend fences, not break them down.

A few suggestions for processing new family information:

1. Take time to think about new information or insights you have.
2. Be self-accepting. Even if you were a volunteer rather than a victim due to certain family events, you too did the best you could do at the time.
3. Empathize. Try to put yourself in other people's shoes. Why did relatives do what they did? Does their history explain their behavior?

4. Blame less, accept more. Fight the urge to criticize or blame relatives for what you feel was unforgivable or unforgettable. Be willing to bury some of the past.

5. Edit new information before disseminating it. Decide what you want to keep to yourself and what you want to share with someone else.

6. Move ahead. Use the new information constructively. Keep things in perspective. Change what you can, don't try to change what you cannot, and understand the difference!

8
GROUP GENOGRAMS:
A GROVE OF TREES

A Genogram Party

If doing a family genogram alone sounds like work, make it play by inviting several close friends to a genogram party. Turn an isolated effort into a comfortable group experience.

Drawing family trees together can be a unique—and fun—social hour or two. Present the idea as a challenge and opportunity to solve family mysteries or get the ghosts out of everyone's closets. The party atmosphere you create is likely to get everyone to approach the task with curiosity and some levity. History doesn't have to be dull. You want to see the humor, the funny quirks in personalities, and crazy coincidences that can make your personal histories amusing as well as interesting. Mixing the heavy with the light allows you to view your family tree from several perspectives. All families have troubles, but they also have good times.

Friends can often lend an important advantage. They

can show you the forest that you cannot see because of the trees. They can help you achieve an overview of your family that you may be too emotionally close to see. The comments and questions of friends can also make you want to dig even deeper into your family roots. Your friends may say things about their families that arouse your curiosity about your own. An added bonus of the group effort is that you and your friends get to know each other better.

A genogram party offers many specific benefits. People in a small group tend to talk out loud as they come across items of interest. This spurs questions and comments from others and can bring out hidden patterns and meanings. It is usually a relief, too, when you learn that other families have their own "ghosts" in their family closets. What you used to think of as "strange" or unique to your family may suddenly seem normal when you find out that your friends had similar experiences.

The spirit of sharing can bring many of the family black sheep, skeletons, and secrets out of the closet. The black sheep in your family are probably not very different from the black sheep in other families. Most families deny or ignore black sheep—people who are the troublemakers or recipients of the family blame. Most families also have heroes—people who are put on a pedestal and honored by the family. Most families have "secrets," too; these are those unmentionable subjects surrounded by mystery, guilt, and shame. Some families keep secrets for generations, and once those secrets are finally exposed, they usually turn out not to be very compelling at all. In fact, often they are "much taboo about nothing."

Generally speaking, families, like individuals, are more alike than different. In talking with your friends about these similarities, you'll find out that even though individuals have different backgrounds, the things that go on in families are pretty much the same. All families have occasions to celebrate and to grieve. Life is a mixture of pain and pleasure, and family life is no exception. Hearing about the trials and tribulations of a friend's family may remind you of crises in your history. Questions you might ask each other are: What were the problems, and what were the solutions? Does your family have a pattern of fighting or taking flight when trouble comes? How did people in your family deal with crisis, and how do you?

Some small groups decide on a specific topic for their genograms. For instance, you might investigate a theme or family message, as discussed in Chapter 5. Or you can make roles the subject of exploration. The roles of family heroes, favorite sons and daughters, rescuers, clowns, troublemakers, and others discussed earlier in this book, are some you might focus on. Or you might concentrate on family disorders, which can be physical or emotional problems like eating or drinking problems, depressive states, or medical conditions like diabetes, heart disease, or cancer.

Constructing genograms in a group is a lively exercise. Undoubtedly your group will conclude that many family trees breathe the same air. The facts may be unique, but the feelings surrounding those facts are truly universal.

Three Friends' Discoveries

Three very good friends—Nina, Molly, and Cindy—spent a recent Sunday evening doing their family genograms together. Earlier that week they had each done some simple research into their families and came prepared with the names of relatives and the years of their births, marriages, and other important events. Each person also brought snacks and "kindergarten supplies." This helped set a lighthearted tone of play time. They gathered around the kitchen table at Cindy's house with all their supplies: sample genogram forms, large sheets of white paper, and an array of colored pencils and pens. They spent some time reading the instructions on how to complete a genogram and then discussed what they wanted to look for as they plotted their families over three generations.

Nina said she would like to figure out her family theme—the main tradition or idea running through her family. Molly said that looking for a theme was interesting to her, too, but she also wanted to learn about family roles. She knew what her family role was (caretaker), but she was curious to see who else played this role in her family. Cindy said she knew there was a "secret" in her family that she had always been curious about. Though she knew what the "secret" was, she wanted to find out why it was forbidden to ask anyone about this taboo subject. The group decided to look for themes, roles, and secrets and to share their individual discoveries as they constructed their genograms.

As you might imagine, they learned a great deal from this exercise. Even more importantly, they had a good time in the process and felt very close to each other. Here are some of the things that went on at the genogram party.

Nina started off with the statement, "My family is not very interesting. Everyone is very prim and proper. I'm afraid my genogram is going to be pretty dull." That was hardly the case. As the evening progressed, all the women commented upon how interesting all of their stories were. There was nothing dull about the repetitive patterns and belief systems that emerged as they recalled and drew their genograms.

The women were surprised at the many feelings that came up for them. Even though doing the genograms together was fun, there were moments of profound sadness. It was painful to think about parents and grandparents who had obviously suffered greatly in their lifetimes. Losing parents was an especially traumatic event in several family histories. The women realized that they were not always objective critics in this process; they were compassionate observers as well.

Another surprise was that the group genogram could be such an intimate experience for all of them. They not only felt closer to the family members they were describing, but they felt closer to each other as they expressed feelings about both the past and present.

Family Togetherness and Abandonment

Cindy decided that one theme running through her family history was togetherness. She related, "Our family is very close. The message my six siblings and I always got was, 'The family sticks together no matter what. The family, right or wrong, but always the family.' I guess you could say that my parents were both rescuers or enablers. They were always there to bail us out. And we did require considerable bailing.

"Most families have one black sheep. But in our family, there were five. I was the only 'goody two-shoes,' and I got no attention at all and certainly no rewards for being good. My four brothers and sister were constantly in some kind of trouble. My parents were always, and still are, racing to the rescue.

"Jerri, my sister, is thirty-five and still lives at home. She has all kinds of emotional problems and is never able to hold down a job for very long. She's the 'sick child' who continues to need my parents' support.

"My brothers have simmered down considerably, but in their teens and twenties they were into everything there was to get into: sex, alcohol, drugs, war, and jail. Not too long ago, my parents posted their home as bail for a brother who got caught trafficking drugs. There was never any question about the lengths my parents would go to save or protect their kids. Their unspoken message was, 'No matter what, we will never abandon our children.'

"This genogram (Diagram 52) makes it very clear where the theme of nonabandonment comes from. Both of my parents were abandoned as children; they were little orphans. My father, in a way, was only half-orphan. His father divorced his mother and left home when my dad was one and a half years old. My father only saw his father once after that, when he came to the college stadium to see his son play football.

"But my mother was certainly an orphan. Her parents had their six children taken away from them and placed in an orphanage. No one knows exactly why. That's the family scandal and secret. My mother was taken to the orphanage right after her first birthday and was adopted when she was three. In those days child welfare authorities separated brothers and sisters, which now seems very cruel. Once she was adopted, my mother never again saw her siblings. They were all placed with different families.

"The strangest thing of all, to me, is that my mother claims to have no curiosity about her natural parents. She

DIAGRAM 52 — ORPHANS AND OTHERS

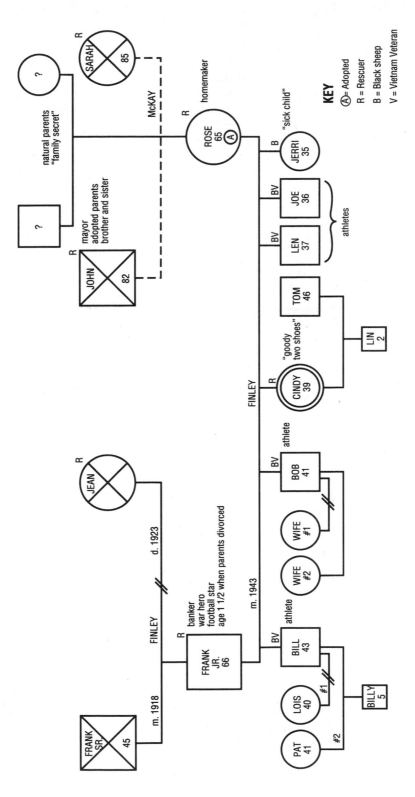

KEY

Ⓐ = Adopted
R = Rescuer
B = Black sheep
V = Vietnam Veteran

only talks about her adoptive 'parents,' the brother and sister who adopted and raised her. We grew up feeling no one was allowed to ask my mother anything about her childhood. It was a taboo subject.

"My parents met on a blind date and got married three months later. Is it just a coincidence that these 'two little orphans' managed to find each other? I wonder. At any rate, they promptly married and promptly had six children. We've been a 'together' family ever since. Everyone has stayed close, and the nest has never really been empty. In fact, our family joke is that once you get into the Finley family, you're never out of it—you're never able to leave. My two oldest brothers divorced and remarried, but their former wives, years later, are still very close to my parents."

Nina was quite moved by Cindy's story. She said, "Cindy, that story is incredible. I know your mother, and she seems like such a free spirit. I had no idea of her devastating history. To think that when she was your daughter Lin's age, two, she was in an orphanage. The idea of this two-year-old separated from both parents and all her brothers and sisters is just awful. How does anyone ever recover from that? Don't you think that's why abandonment is such a big issue for her?"

Cindy responded, "I used to be very angry at my parents for being so overprotective and intrusive. In this light, though, I'm beginning to see what loss means to them. Before, it never made much sense to me why no one was allowed to leave the nest. But neither of my parents really ever had the safety of a nest. They were shoved out much too soon. No wonder it's so important to them. Their experiences of abandonment, even though they never talked in those terms, has to account for why they cannot think of abandoning anyone close to them. And I see why the nest they made together has never been empty. One or more of their adult grandchildren has always lived with them."

Roles of Responsibility and Irresponsibility

Cindy continued, "Another pattern in my family is that

the men are heroes, and the women take care of the men. If I heard it once, I heard it a thousand times: 'Why don't you see if your brothers want (whatever)?' Females are expected to wait on the males.

"My father was a college football star and a much-decorated officer in World War II. All four of my brothers served in the Vietnam War. They weren't exactly heroes, but they volunteered to serve in a very unpopular war. They complied with the family rule that said, 'Men are athletes and soldiers.'

"I'm a caretaker, and from my genogram it's not hard to see where I got those skills. My mother was a superwife, supermother, and superdaughter. She took care of my father, six children, and her invalid (adoptive) mother, who lived with us for twenty years. I certainly never remember my mother sitting down very often. She was always on the go, doing something for someone all the time.

"Her adoptive mother was a caretaker too. She never married and from the age of sixteen kept house for my grandfather (actually her oldest brother), who was a bachelor and mayor of the New England town where the family had lived for several generations. He was a pillar of the community, much older than she, and I don't suppose he helped out at home."

Family Rules

When Cindy was discussing the rescuers and black sheep in her family, Nina was reminded of the similar roles played out in her family (Diagram 53). "We had a whole different background. My parents were midwestern WASPS, and I certainly never heard about any scandals or even improprieties. But we also had our share of rescuers and black sheep. However, the black sheep were called, if anything, 'relatives who were having a bad time.'

"The theme in my family was, 'Don't ever raise your voice.' I don't think my parents *ever* had a fight, and I certainly have trouble having them. I often wish I could just haul off and yell and scream when I'm angry at my husband or kids. But, like my parents, all I show is a stern look of disapproval. And even though I know how guilty and terrible those kinds of looks can make people feel, I end up putting on that same face whenever I'm angry. I'm constantly working on saying what I feel when I feel it. My feelings don't come out easily for me. I see that they never came out for my parents.

"My parents were certainly rescuers. They had the reputation in the family for being the level-headed people that anyone with a problem would come to for help. I was their only child and didn't require a tremendous amount of rescuing. Still, I always knew they would be more than happy to rescue me. But my aunts, uncles, cousins, and grandparents were frequently the victims that my folks had to save. I might say they did a great job at it. Unfortunately, I've found I have a tendency to do the same thing. My husband George, like many of the men in my life, often plays the 'victim.'

"For years my parents had tried to have a baby and, after three miscarriages, they finally had me when my mother was thirty-five years old. I guess I was kind of a miracle for them, and I could do no wrong, or so it seemed. Growing up, I interpreted their low expectations of me as meaning they didn't think I could excel in anything. I felt that whatever I did was less than mediocre but OK with them. I knew they would save me from situations I could not handle on my own."

At this point, Molly interjected, "But Nina, can't you see that it wasn't that they had low expectations of you, but that they may have been totally accepting of you just the way you were? They probably had almost given up hope of having a family when you were born. The very fact of your being, and especially that you were healthy, was enough for them. I would guess that they were so relieved and happy to have you that you were perfect in their eyes and didn't have to do anything to make them happy. You were quite enough."

Nina replied, "When you put it that way, it does make sense, but still I cannot erase the fact that for years my feelings were that they didn't have that much confidence in me. Perhaps that was simply my self-induced low self-esteem. I'm sure they would agree with you if asked. But,

DIAGRAM 53 — VICTIMS AND RESCUERS

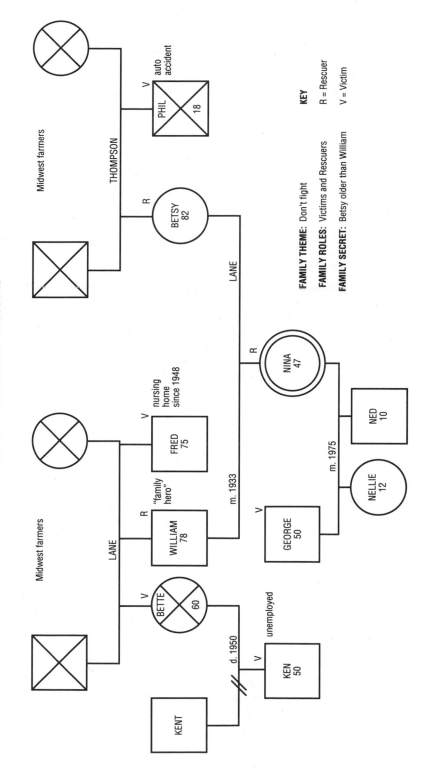

KEY

R = Rescuer

V = Victim

FAMILY THEME: Don't fight

FAMILY ROLES: Victims and Rescuers

FAMILY SECRET: Betsy older than William

for whatever reason, I did not experience their lack of criticism as acceptance. I ended up feeling I was not good enough even if they meant to imply I was fine the way I was.

"At least now I have some real sympathy for parenting. It sure seems as if you cannot win. You cannot guarantee your children's success and self-confidence no matter what you do or don't do. And now that I have two children, I see how you cannot do everything right. My kids misread me too."

Comments and Comparisons

Molly was the last person to discuss her genogram. She said, "I loved hearing about both of your families. It is interesting how different yet alike our families are. Cindy had a Boston Irish story, and Nina's was a midwestern WASP story. Mine is a first-generation Russian Jewish story. Still, I'm struck by the profound early childhood losses in Cindy's family and the rescuing roles in Nina's genogram. Both of these factors were certainly significant issues in my family.

"My father lost a parent at a very young age also. He was twelve when his father dropped dead of a heart attack right before his eyes. One minute his father was alive and well, and the next he was dead and gone forever.

"I always felt terribly sad for my father that this happened to him, but it really is only this evening that I am beginning to understand how that tragedy negatively shaped my father's world. It had to affect all of his experiences after that. It accounts, too, for the family theme of 'black and white, good or evil, all or nothing.' The world was basically perceived as a hostile place.

"I used to accuse my father and his brothers of being rigid and arrogant. But now all that makes some sense. How could you not see things in such absolutes like here today, gone tomorrow, if that's exactly what happened to you? They were very skeptical—people were for or against them.

"My father quickly, and involuntarily, became a rescuer. He was the oldest of four boys and had to quit school to help support the family. His mother took in boarders, but his income from his menial job at the local printing plant was absolutely necessary to keep the family going.

"Ironically, my father, who saved the family, always felt like a failure in comparison to his brothers, who finished school and did very, very well in their careers. They were rich, and he always said he was 'poor,' even though he was a steady worker and a good provider. In his eyes, he didn't measure up to his college-educated, sophisticated brothers. He 'earned wages' while they 'made money' at their own businesses.

"I used to be surprised that my father's brothers were just like him. But, of course, that makes sense. He was the man of the family after his father's death. Even though he was only a few years older, they must have looked up to him and used him as a role model. He saw the world in a limited two-dimensional way, and so then did they. No one taught or convinced my father that the world wasn't a very cruel place. He didn't feel he could let down his guard long enough to consider that people may not be all good or all bad. He didn't have time to consider subtleties or complexities.

"There were rescuers in my mother's family, too, but they were all women. That rescuing usually took the form of caretaking. My mother was a supercaretaker, and she had four daughters. Three of us are in caretaking professions (nursing and teaching), and the fourth is the black sheep, the baby who is still crying out to be rescued.

"The ultimate example of supercaretaking is my sister Joan. She is married and has three children of her own. For the past ten years, she has also raised our baby sister's two children, one of whom has emotional problems. Sue is a drug addict and has never been able to take care of herself or her children. She literally dumped her kids at Joan's house when they were toddlers, and they've been there ever since. Joan is loving and supportive and has raised them as her own. Both she and her husband have been wonderful to the kids.

"Joan and Sue are examples of the all-good or all-bad roles that conform to my father's black-or-white world. You are either responsible or totally irresponsible in our

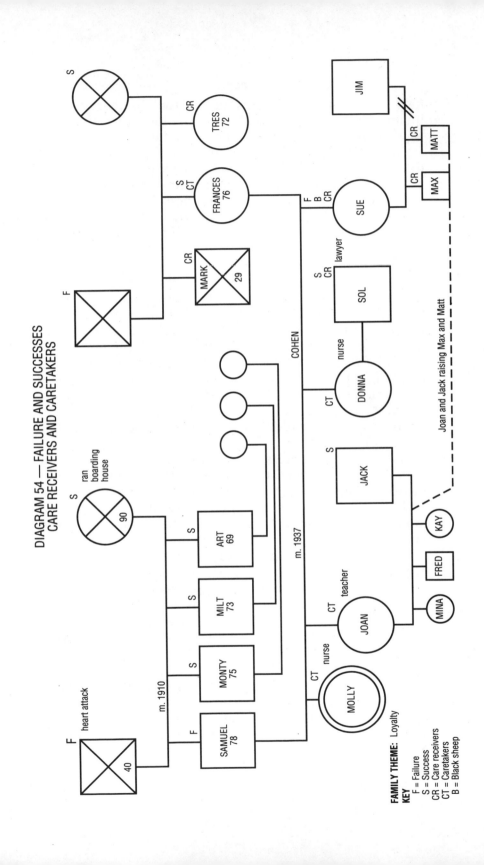

DIAGRAM 54 — FAILURE AND SUCCESSES
CARE RECEIVERS AND CARETAKERS

FAMILY THEME: Loyalty

KEY
F = Failure
S = Success
CR = Care receivers
CT = Caretakers
B = Black sheep

family. My genogram can be divided into two distinct camps: the people who took care and the people who needed the care."

The Genogram Challenge

These are some of the highlights from one group genogram session. You can study these three genograms and compare them with your own. After doing their genograms together, Cindy, Molly, and Nina continue to discuss connections as they further explore their family trees on their own. They were so enthusiastic about the genogram party that they recommended it to other friends and even a few family members.

Whether you draw your genogram alone or with friends, it is bound to be a challenge and a special event—a truly enriching and enlightening activity. Enjoy the experience.

BIBLIOGRAPHY

American Genealogical Research Institute Staff. *How to Trace Your Family Tree.* New York: Dolphin Books, Doubleday & Co., 1974.

Bowen, Murray. *Family Therapy in Clinical Practice.* New York: Jason Aronson, 1978.

Carter, Elizabeth A. and McGoldrick, Monica. *The Family Life Cycle: A Framework for Family Therapy.* New York: Gardner Press, Inc., 1980.

Lehman, Kevin. *The Birth Order Book: Why You Are the Way You Are.* Old Tappan, N.J.: Fleming H. Revell Co., 1985.

McGoldrick, Monica and Gerson, Randy. *Genograms in Family Assessment.* New York: W.W. Norton & Company, 1985.

Toman, Walter. *Family Constellation.* New York: Springer Publishing Co., 1976.

Zimmerman, William. *How to Tape Instant Oral Biographies.* New York: Guarionex Press, Ltd., 1979.

INDEX OF DIAGRAMS

INDEX

Abuse, 21
Accidents, 114
Adoption, 6–7
Age, 45–46, *illus.* 46
Alcohol abuse, 29, 31, 61, 100, 102, 126, *illus.* 30, 124
"Always on the outside" triangles, 72–73
American Genealogical Research Institute, 4
American Genealogical Society, 24
Anniversary reactions, 114–15

Basics genogram, 41, *illus.* 49
    format and symbols for, 43–48
    step-by-step construction, 44–48
Bird's eye view, of family, 35–36
Birth dates, 45–46, *illus.* 46
Birth order, 110–11, 113–14, *illus.* 112
Birth Order Book: Why You Are the Way You Are (Leman), 111
Blackboard, use of, to draw first draft of genogram, 22
Bowen, Murray, 3–4, 75

Calm period, drawing genogram during, 17–18
Carter, Betty, 4
Changes, making, 128–29
Coincidences, 36, 61
    anniversary reactions, 114–15
    birth order, 110–11, 113–14, *illus.* 112
    events, 109–10
    separations, 115, 117, *illus.* 116
    use of genograms to uncover, 12–14
Crisis, drawing genogram during, 18–19

Dance of Anger (Lerner), 79
Death, 32, *illus.* 46
Decision, to vacate triangle, 80, 81
Defusion, to vacate triangle, 80, 81
Detachment, to vacate triangle, 80

Details genogram, 42–43
    preparing, 52–54, 57–59, *illus.* 55
    repetitive patterns in, 61, 63, *illus.* 62
Distance, to vacate triangle, 80–82
Distances genogram, 41–42
    symbols for, 50–52
Divorce, 45, *illus.* 45
Documents, use of, to draw genograms, 24–25

Events, power of, 109–10
Extended family genograms, 63–64, 66, *illus.* 65

Family, use of genograms to understand, 8–12
Family lore, *illus.* 109
Family myths, untangling, 102–4
Family research, importance of, in researching genogram, 28–29, 34
Family scripts, use of genograms to uncover, 5, 14, 94–95, 97, 99, *illus.* 96
Family stories, extending genograms to include, 1
Family systems theory, 4
Family traditions, identifying, 83–85, 87, 89, 91, 93–95, 97, 99–100, 102–6, *illustrations* 86, 88, 90, 92, 96, 98, 101, 105, 107
Family therapists, use of genograms by, 4
Family trees, 3, 11
    adaptation to include client history, 3
    bringing, to life, 31–32, 34
    talking, 15–16
Fights, identifying in families, 89, 91, *illus.* 90
Folklore, *illus.* 107
Fun, making genograms as, 4–8

Gender symbols, 32, 43
Genealogy, 4
Genogram(s)
    basics, 41, 43–48, *illus.* 49

151